BAT ISLAND

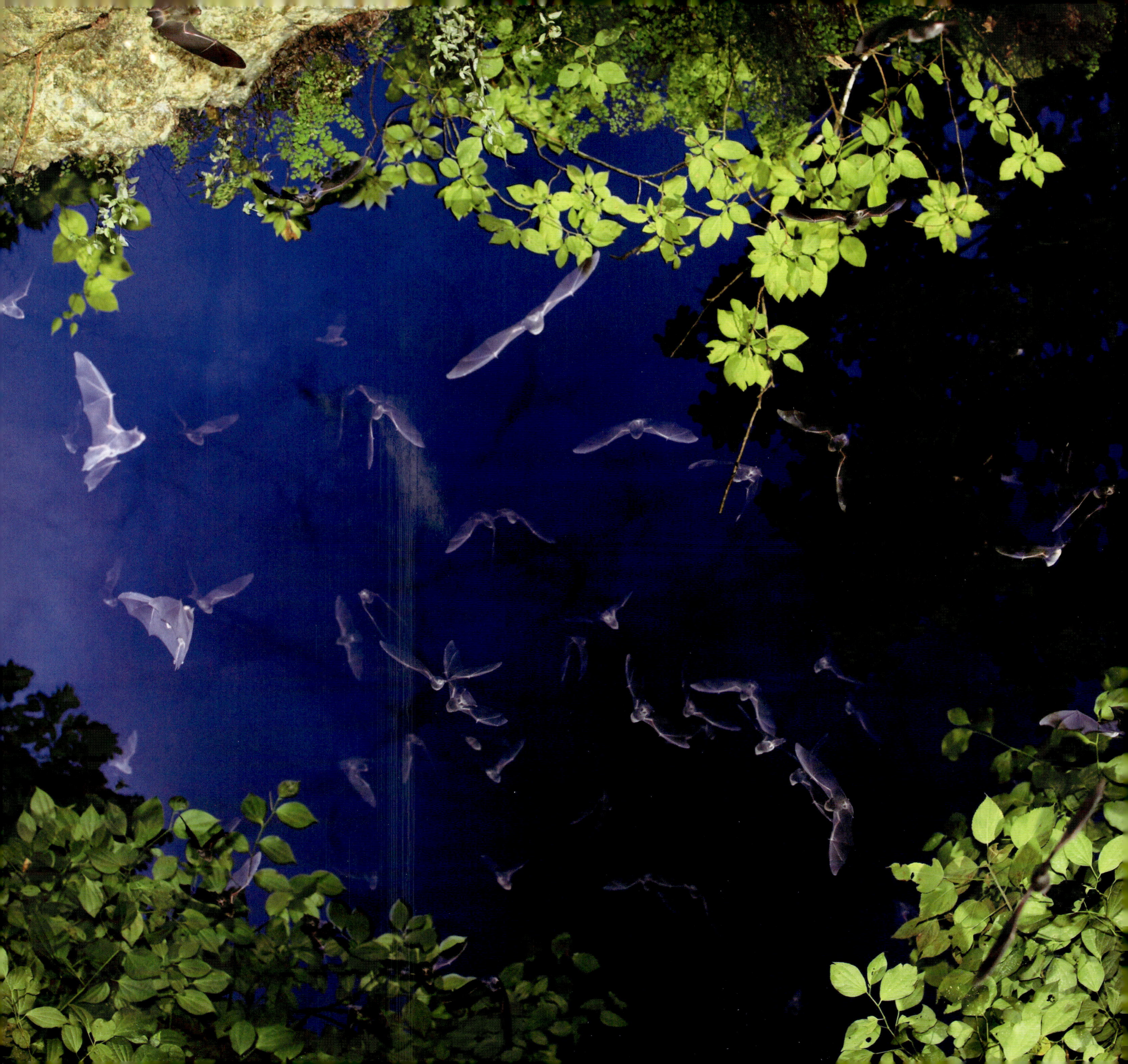

PHOTOGRAPHY BY CHRISTIAN ZIEGLER

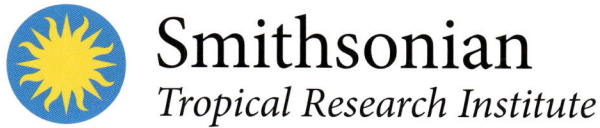

Smithsonian
Tropical Research Institute

BAT ISLAND

A Rare Journey into the Hidden World of Tropical Bats

RACHEL A. PAGE • DINA K.N. DECHMANN • M. TEAGUE O'MARA • MARCO TSCHAPKA

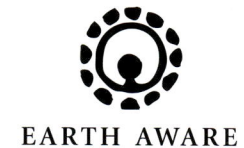

EARTH AWARE

We dedicate this work to the extraordinary legacy of Elisabeth Kalko. In Eli's honor, the Smithsonian Tropical Research Institute has created a memorial fellowship to support the next generation of Neotropical bat researchers. Proceeds from this book go directly to this fund.

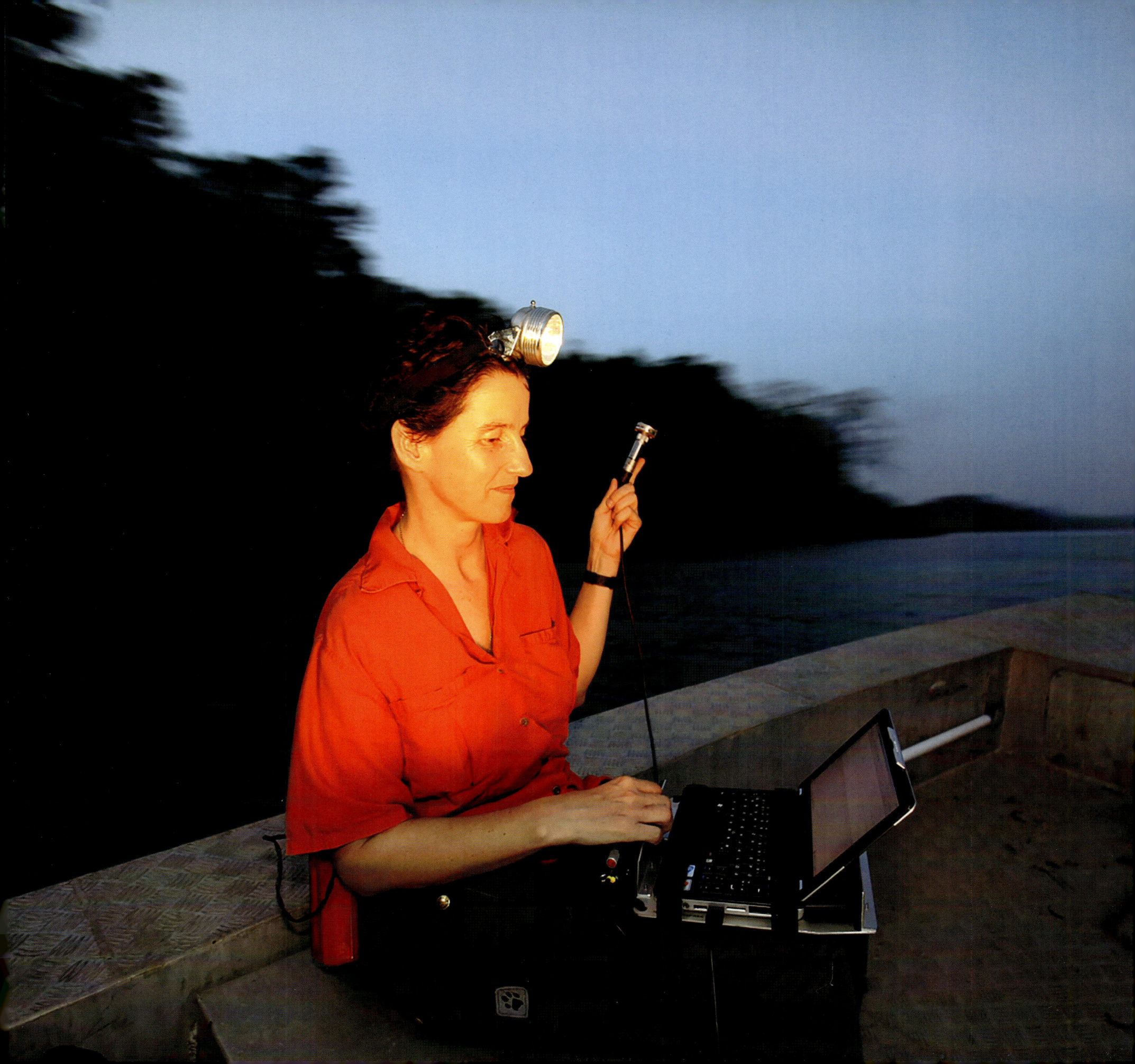

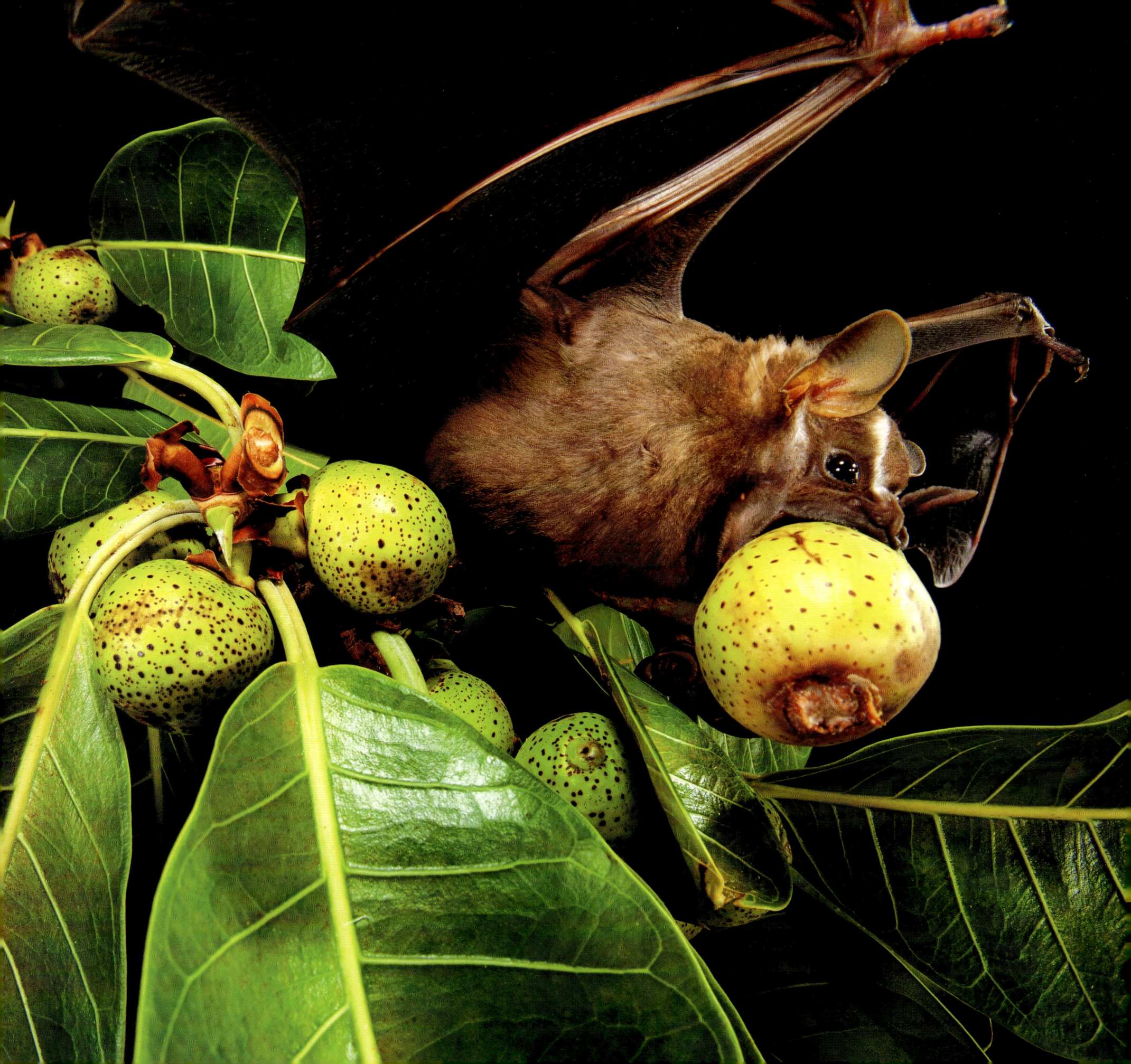

CONTENTS

INTRODUCTION

Years ago, bat researcher Elisabeth "Eli" Kalko and nature photographer Christian Ziegler put their heads together and hatched the idea for a book about tropical bats. Eli and her students had been studying bats in Panama for decades, often accompanied by Christian. They felt it was time to synthesize the unique wealth of knowledge about tropical bats and show the world how diverse and fascinating bats are through Christian's pictures. However, Eli's bright flame burned far too quickly, and in 2011 she passed away unexpectedly during a field trip to Kilimanjaro at the age of forty-nine. Her legacy continues to this day. Panama continues to be a hot spot for bat researchers, many of them her former students, and Christian still documents their work with his breathtaking pictures. And so, many years later, we—Drs. Rachel A. Page, Dina K.N. Dechmann, M. Teague O'Mara, and Marco Tschapka—have come together to fulfill Eli's dream and finally invite the world to see the bats of Panama through Christian's lens.

Bats are extraordinary. With over 1,450 species, they are a highly successful group that make up over 20 percent of all mammal species. They are more ecologically diverse than any other group of mammals, feeding on the broadest range of diets and inhabiting an extensive variety of niches, which they access using a fascinating array of sensory modalities. Even though the order of rodents is higher in species richness, bats surpass rodents in their stunning ecological diversity and their impressive adaptations in morphology (anatomical form and structure), physiology, and behavior. They are also the only mammals to have acquired true powered flight. The key adaptations of bats—flight and echolocation—have allowed for extraordinary evolution and rapid diversification of species. Bats occur almost everywhere on Earth except in extreme deserts, high mountaintops, small oceanic islands, and polar regions. Their species richness is highest in the tropics, where over a hundred species can coexist in a single location.

Because of their large collective biomass and diverse feeding habits, which include plants (fruit, nectar, and occasionally leaves and bark), a wide variety of animal prey (from tiny insects to small vertebrates), and even blood, bats provide critical ecosystem services. The vast majority of bat species are voracious predators of insects and are critically important for the control of insect pests, saving farmers an estimated 23 billion dollars in pesticide use each year in the United States alone. Bats with vegetarian diets are essential for the reproduction of many plants, as they move seeds and pollen over larger distances than most other animals—making them the secret gardeners of many landscapes. Without bats, the regeneration of tropical forests

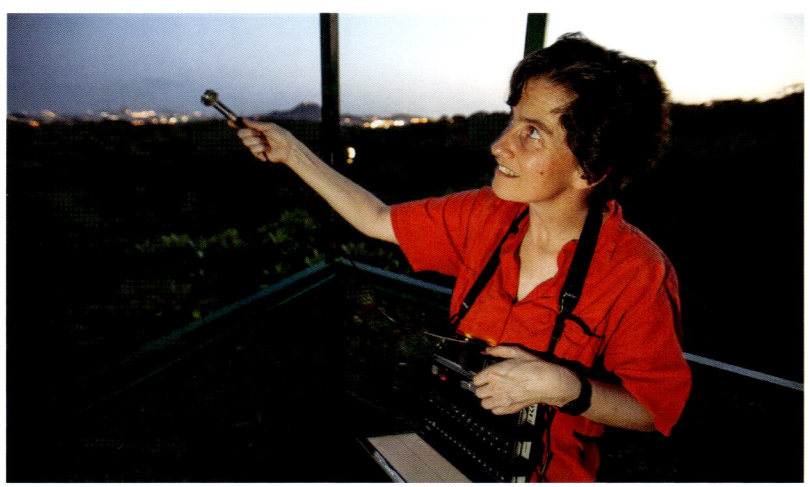

Smithsonian Staff Scientist Elisabeth "Eli" Kalko recording bat echolocation calls from a tower high above the rainforest canopy in Panama. Eli was a pioneer of tropical bat research, and her legacy continues to this day.

would be severely delayed, or in some cases impossible. Through their interactions with a wide range of other organisms, bats are essential in their contribution to the long-term maintenance of diversity.

Bats are fascinating, but studying them is difficult. Most species are very small, highly mobile, and active at night. Bats are thus one of the most poorly understood groups of animals. However, rapid technological advances have revolutionized bat studies in recent years. Innovative new tools including miniaturized tracking technology, thermal cameras, and ultrasound recording equipment have enabled us to study bats in unprecedented detail. One island in the Panama Canal has played, and continues to play, an important role in the groundbreaking studies of tropical bats: Barro Colorado Island. This book is about the bats of Barro Colorado Island and how their in-depth study has changed our general understanding of this extraordinary group of animals.

This book is timely, as bats are increasingly under threat. While many animal species are at risk due to habitat loss and climate change, bats are uniquely vulnerable. On the one hand, they are traditionally perceived as negative in many cultures. On the other, in today's era of emerging diseases, bats are frequently targeted as culprits—often incorrectly, but sometimes correctly, based on solid scientific evidence. To date, there is no conclusive support for direct transmission of SARS-CoV2 from bats to humans. Since the outbreak of Covid-19, however, many studies have focused on the links between human encroachment on wildlife habitat and increased risk of pathogen spillover. Two of the most famous and conclusive cases of spillover from bats to livestock and to humans illustrate this link. The first, in Southeast Asia, is the Nipah virus

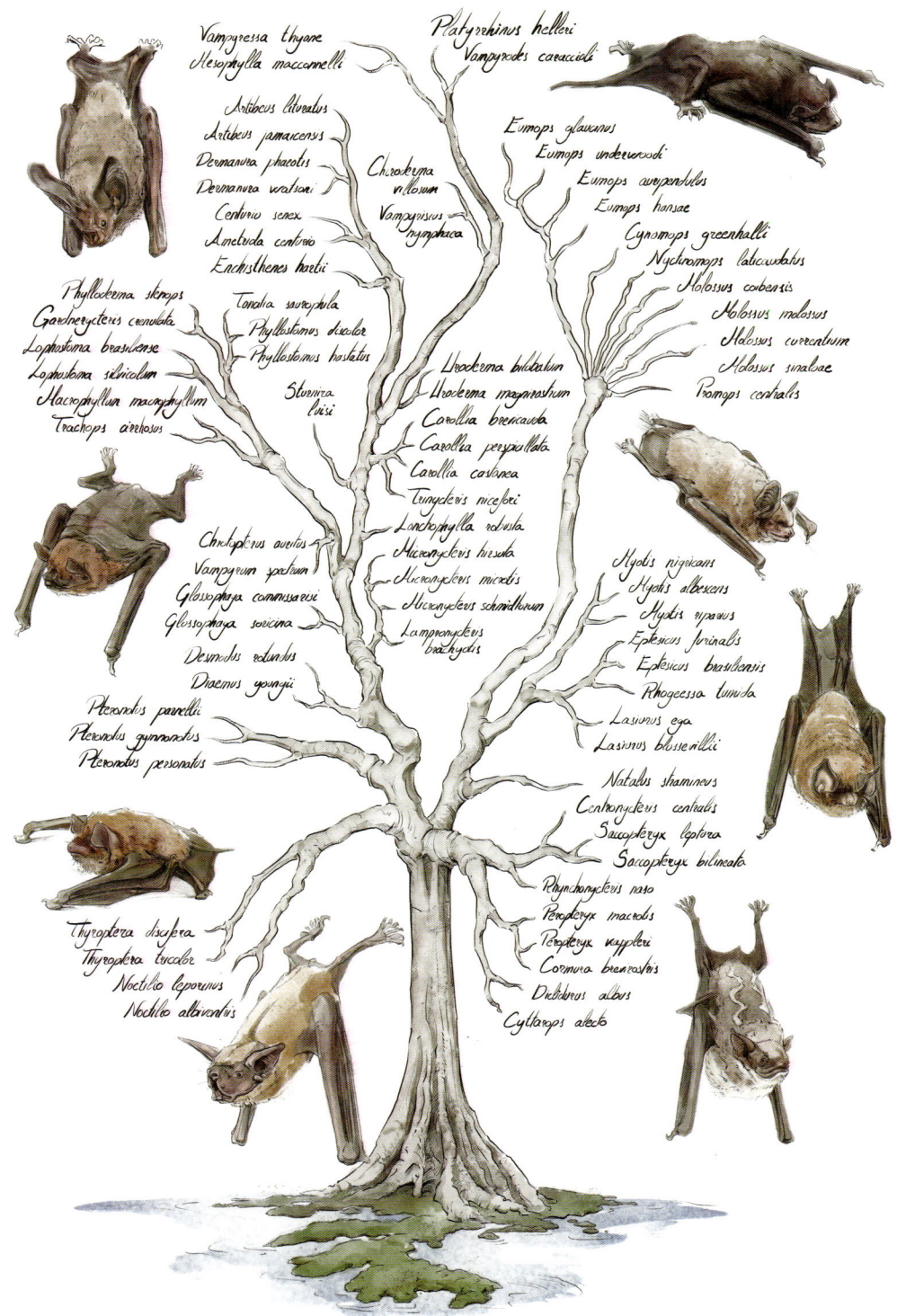

A tree illustrating the evolutionary relationships of the seventy-six bat species from eight taxonomic families found on Barro Colorado Island, Panama. The bat species illustrated are (clockwise from the top): Pallas's mastiff bat (*Molossus molossus*), black myotis (*Myotis nigricans*), Mexican funnel-eared bat (*Natalus stramineus*), greater sac-winged bat (*Saccopteryx bilineata*), greater bulldog bat (*Noctilio leporinus*), Spix's disk-winged bat (*Thyroptera tricolor*), big naked-backed bat (*Pteronotus gymnonotus*), and white-throated round-eared bat (*Lophostoma silvicolum*). Illustration © Javier Lázaro.

that was transmitted through the body fluids of bats roosting in trees to young pigs kept in pens under those trees; the virus then spread to humans as the pigs were sold and transported across the region. The second is the Marburg virus, which was transmitted to Ugandan mine workers through seasonal surges of virus loads from bats in the natural cave system where the bats were roosting. In both cases, spillover could have been avoided through proper understanding of bat biology and behavior. Fear of disease has led to active persecution of bats all over the world.

There is a long history of bat research in Europe and North America, but the study of tropical bats is relatively recent. The Smithsonian Tropical Research Institute (STRI) was one of the first institutions to study tropical bats in detail, and this research initially took place almost exclusively on Barro Colorado Island (BCI). The island now harbors one of the best-studied rainforests in the world. The seventy-six bat species of BCI form one of the best-studied communities of bats in the tropics, and maybe even worldwide. BCI is an amazing natural laboratory, where for a century unique datasets have been gathered that provide crucial insights into the composition, dynamics, and functioning of this species-rich tropical ecosystem. In the following pages, we will look at the many discoveries made on BCI and in Panama over the past decades, place them into a broader context, and highlight impressive technological advances that have opened doors in the study of bats.

This book is a homage to the bats of BCI as seen through the superb photographs of Christian Ziegler, a nature photographer who learned his craft in the forests of BCI. Each of Christian's images is accompanied by brief text that addresses the deceptively simple, yet fundamental, question: "How can so many species coexist in such a small place?" This book explores the different adaptations and strategies that each bat species has used to find its place in the richly structured tropical rainforest. From their over 55-million-year-old ancestors to the more than 1,450 species of modern bats, evolution has crafted a magnificent diversity of species and lifestyles. Bats are obviously a success story, which we will share.

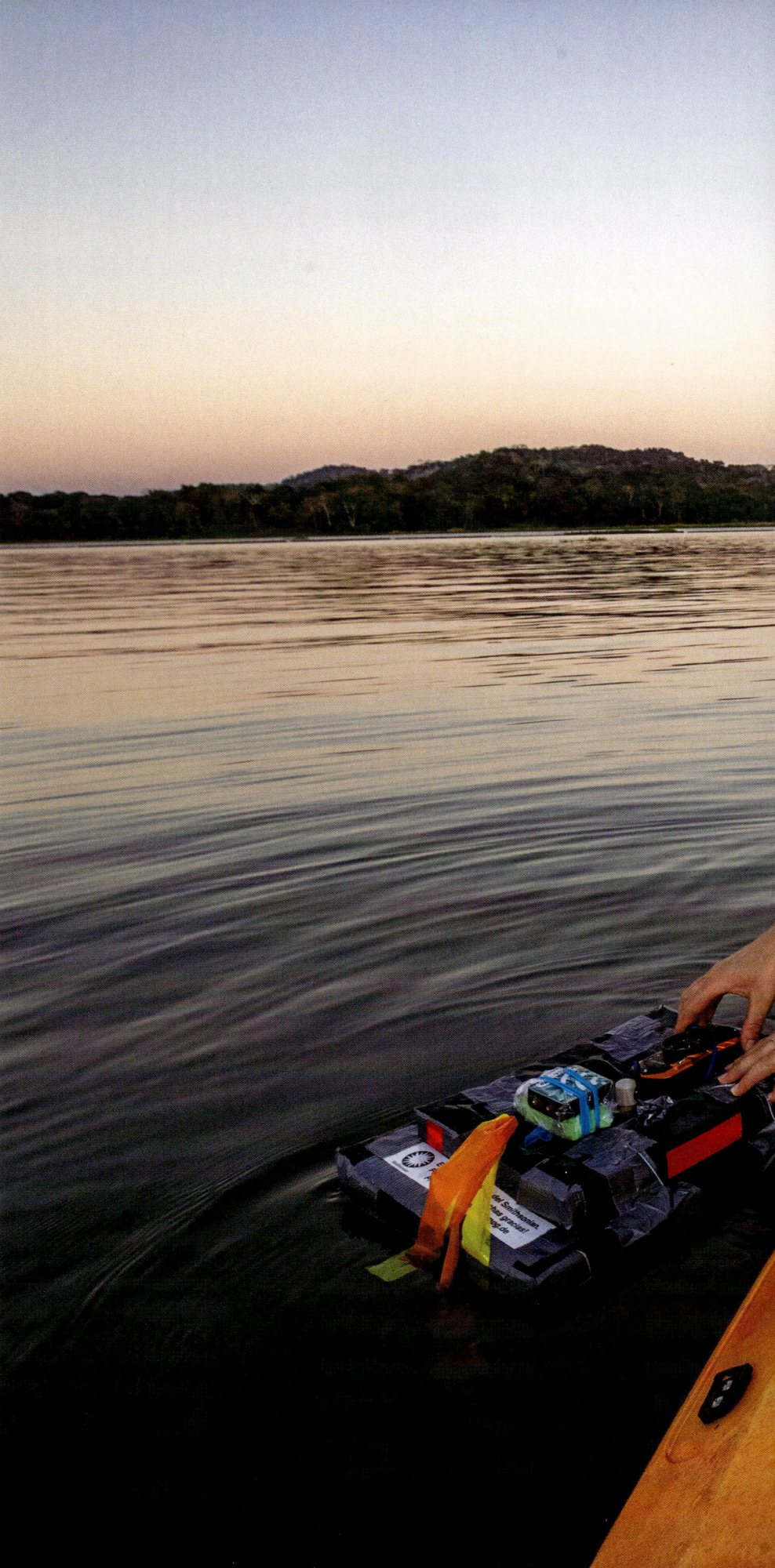

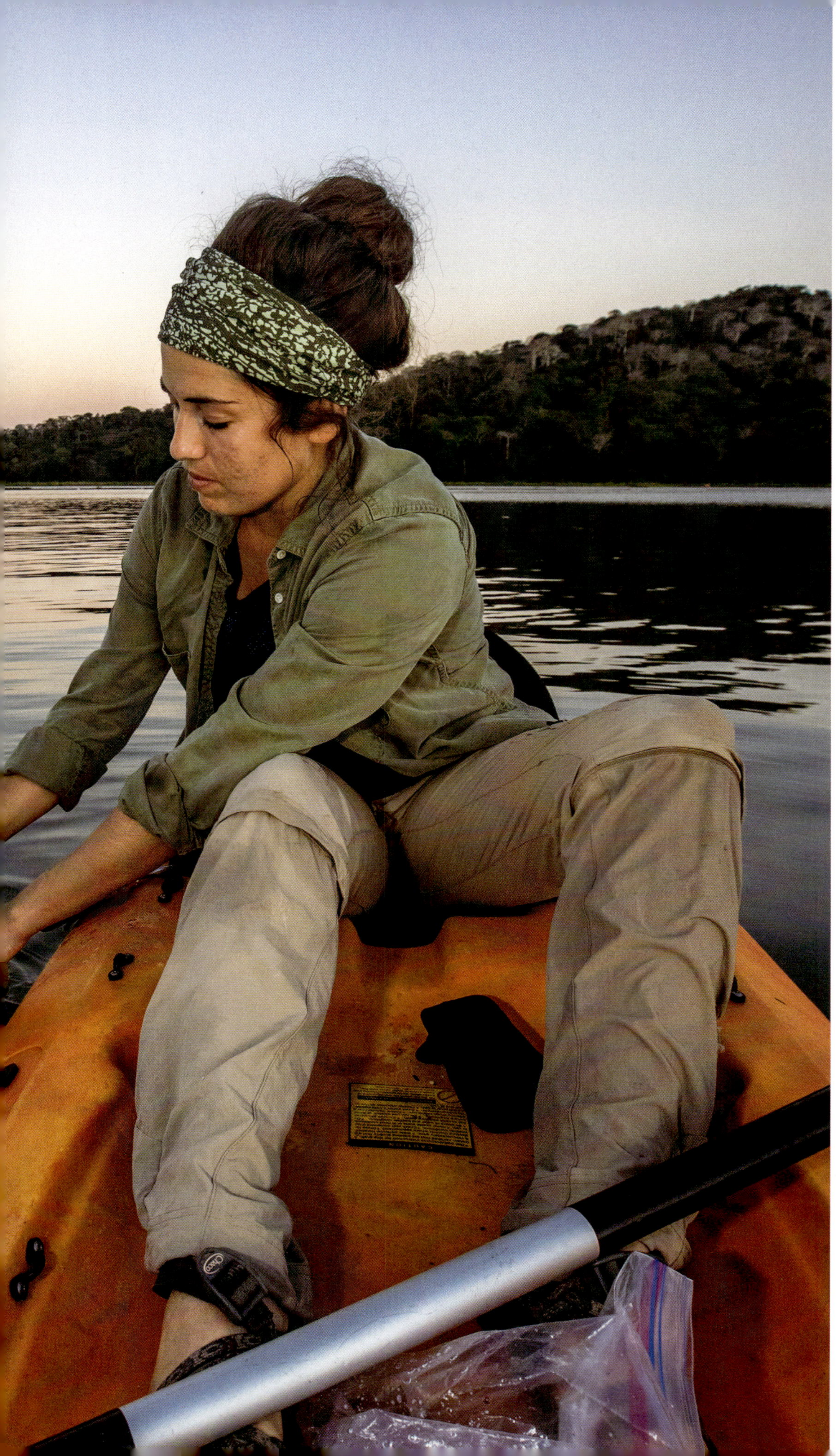

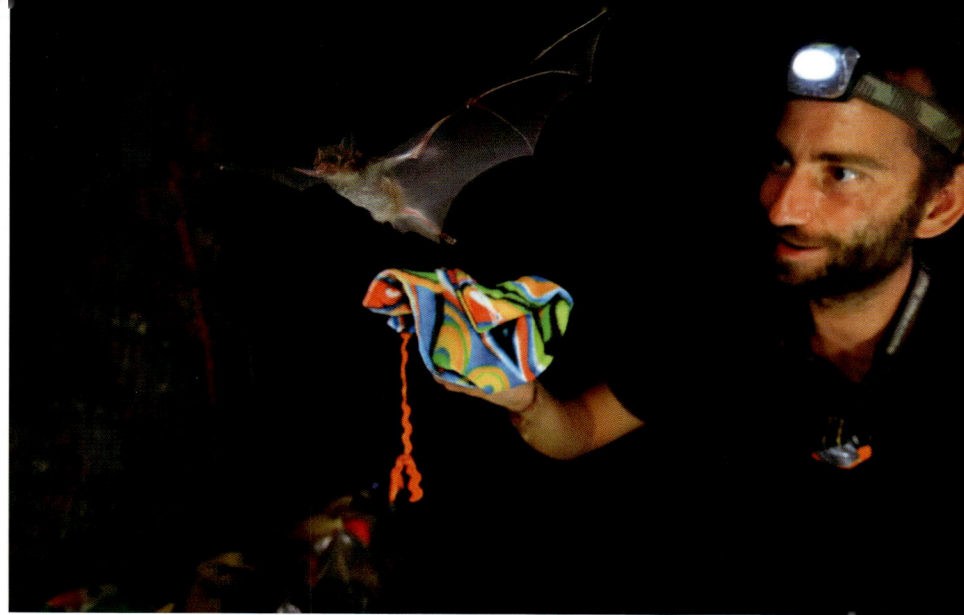
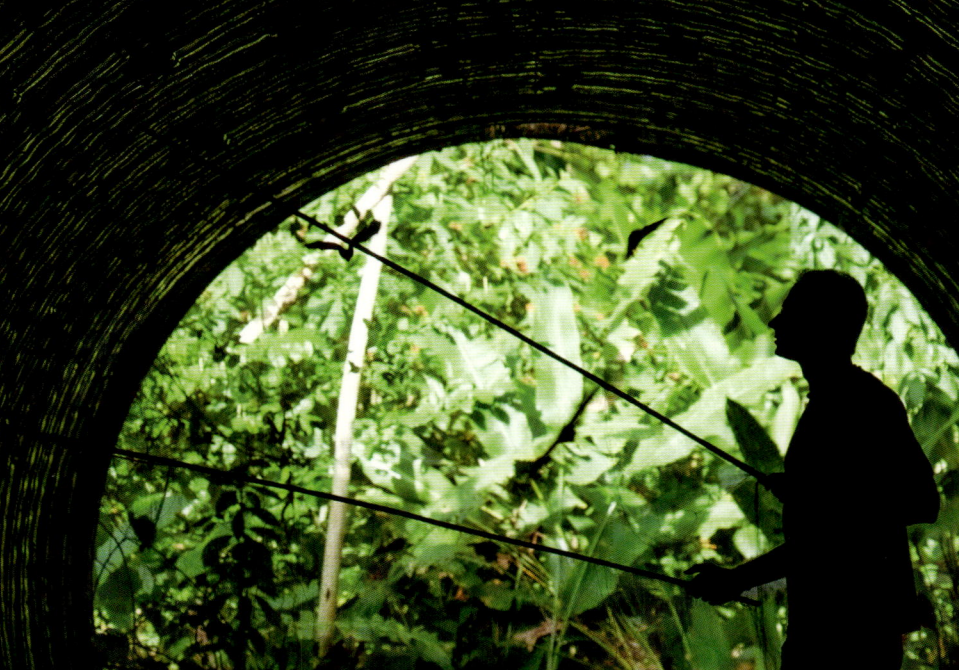
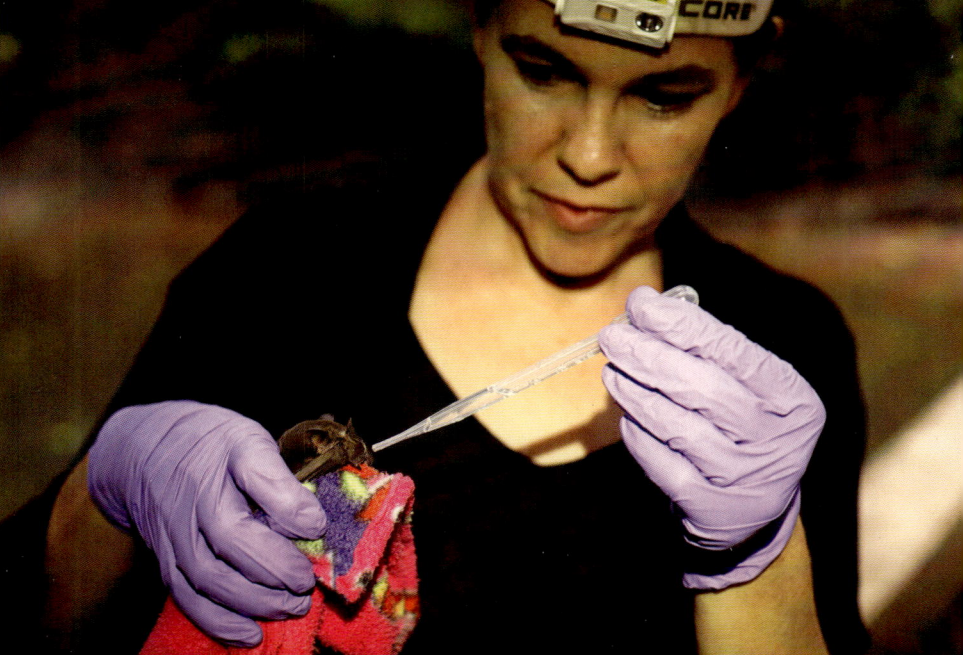

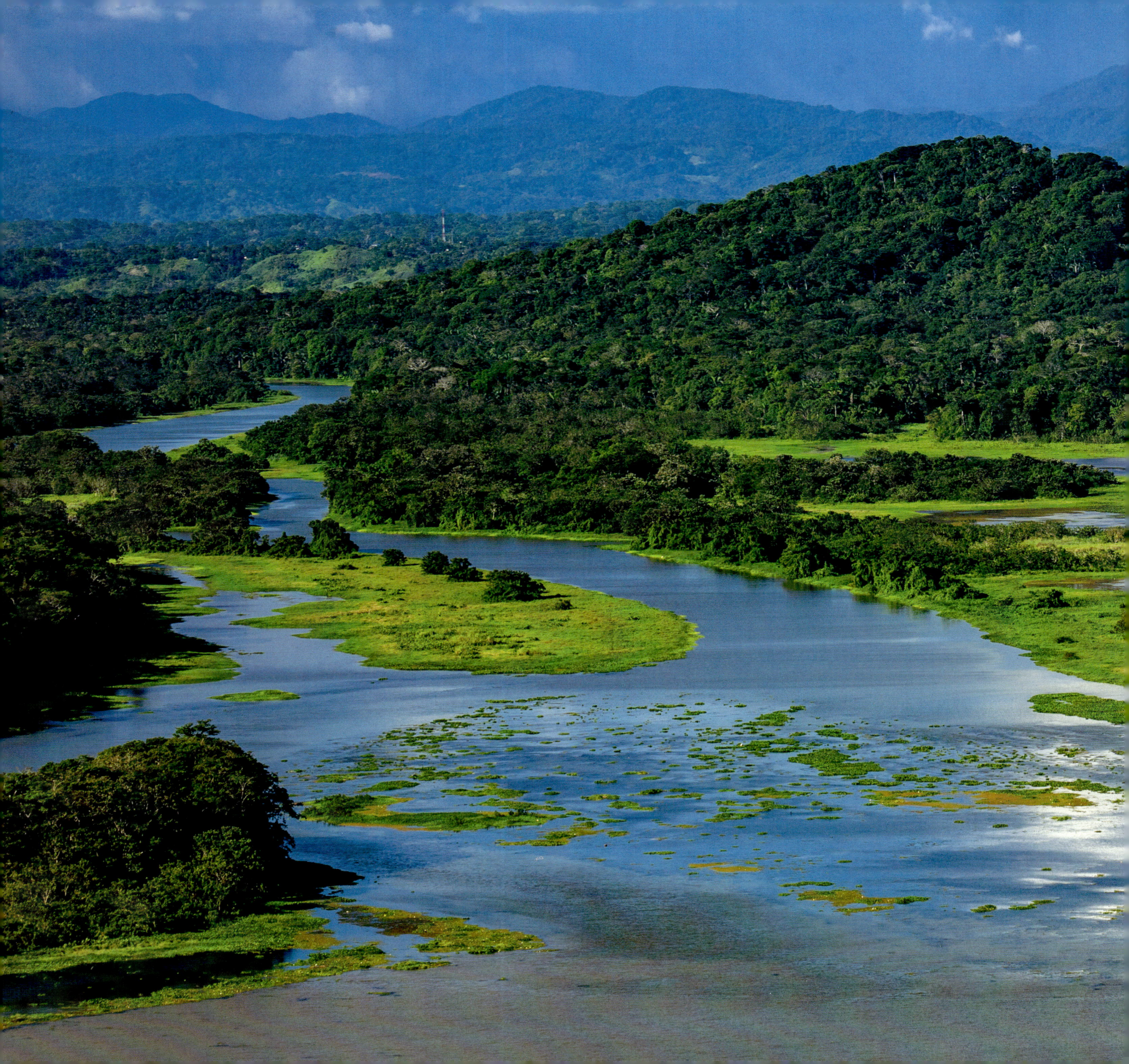

CHAPTER ONE

ISLAND UNDER THE MOON

THE HISTORY OF BAT RESEARCH ON BARRO COLORADO ISLAND

Founded in 1846 and based in Washington, D.C., the Smithsonian Institution is the world's largest complex of museums, research centers, and educational facilities. Smithsonian science in Panama started in the early 1900s, when the United States was completing construction of the Panama Canal. At this time, Smithsonian scientists coordinated a biological survey in the areas around the canal construction zone. The Chagres River was dammed in 1913 to create Lake Gatún, which forms a large part of the Panama Canal. The flooding of Lake Gatún created Barro Colorado Island, the largest island in this newly formed lake, from a former hilltop. The Panama Canal officially opened in 1914, and nearly a decade later, in 1923, Barro Colorado Island (BCI) was proclaimed a nature

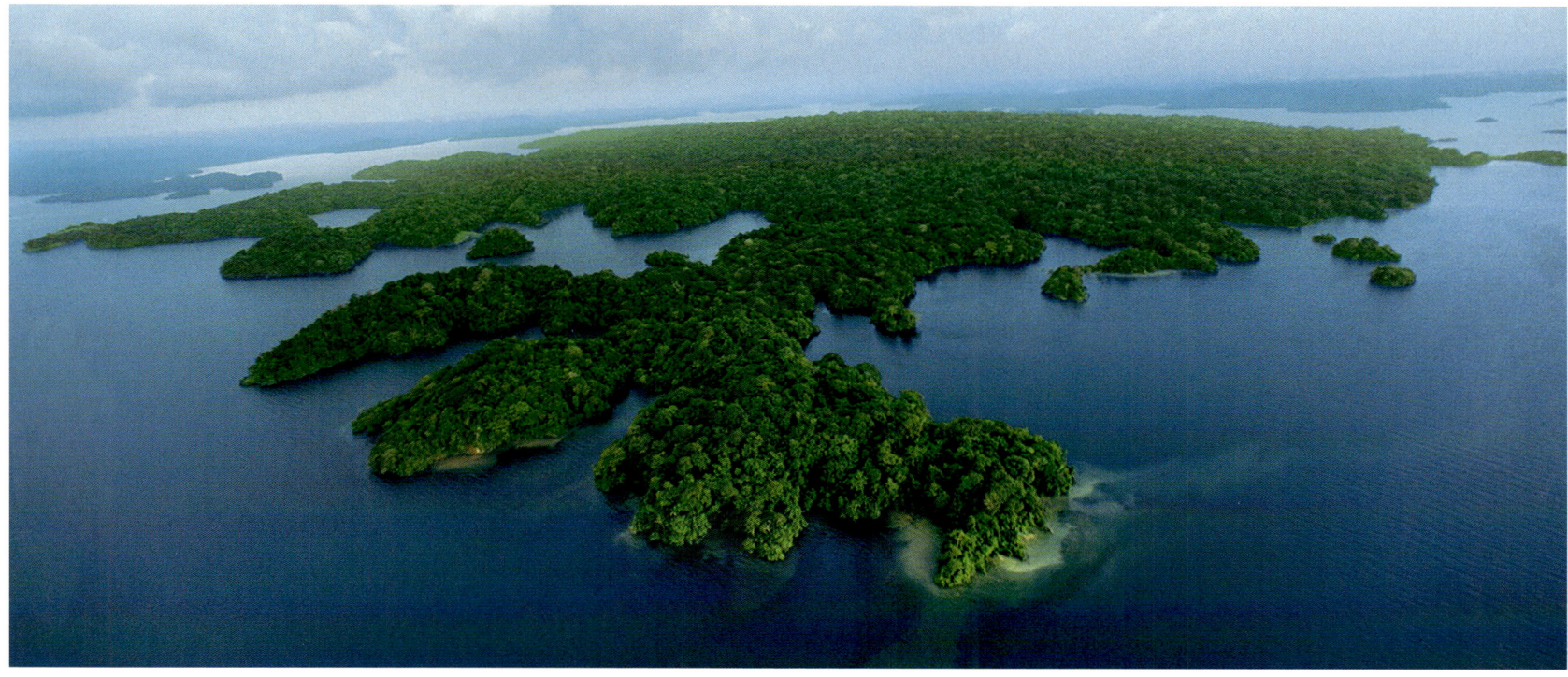

OPPOSITE The mouth of the Chagres River that was dammed in 1913 to create Lake Gatún, turning a former hilltop into Barro Colorado Island. ABOVE An aerial view of Barro Colorado Island, the largest island in the Panama Canal.

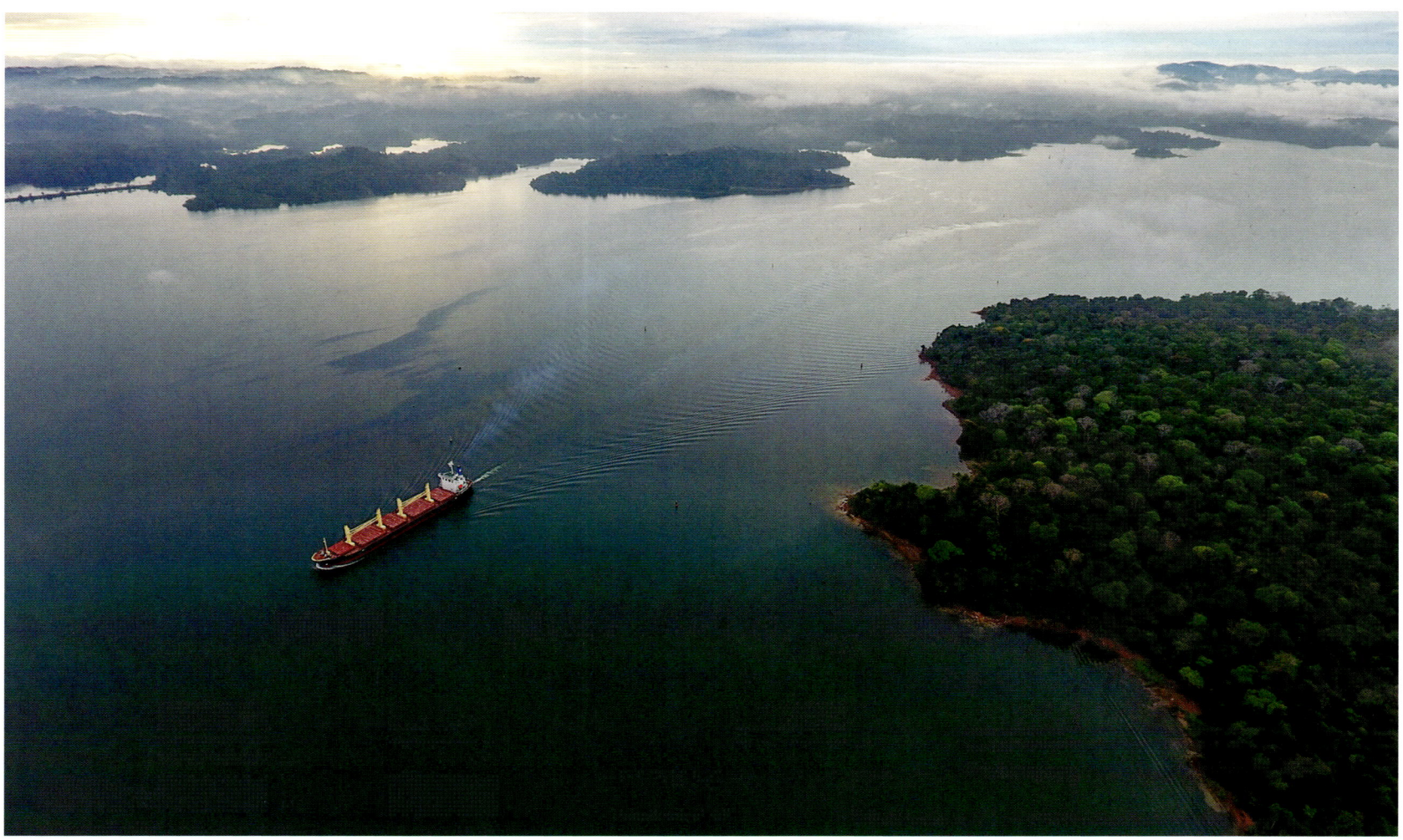

reserve. A research station opened on the island in 1924. BCI officially became part of the Smithsonian in 1946. Now BCI is one of over a dozen field stations spread across the isthmus of Panama and administered by the Smithsonian Tropical Research Institute (STRI), the only Smithsonian unit located outside of the United States.

Barro Colorado Island was an ideal location for long-term tropical research for a number of reasons. The government of Panama was in support of the research station, and the area surrounding the Panama Canal offered infrastructure that facilitated transportation and research. The small field station offered researchers a chance to study the tropical forest at their doorstep. A century later, the body of knowledge that has emerged from extensive, long-term study of this 6-square-mile (15.6-square-kilometer) island is staggering. Each study builds on those before it, offering an unparalleled understanding of the complex web of tropical life. This is true for all taxa studied on BCI, and certainly so for bats.

Worldwide there are over 1,450 species in the order of bats (Chiroptera) and these are divided into twenty-one scientific families. Roughly 120 species have been documented in Panama. Thus, over 8 percent of all the world's bat species are

A ship transiting the Panama Canal passes by Barro Colorado Island. Scientists have conducted research on BCI for over one hundred years.

found on this small strip of land connecting North and South America. For comparison, in the United States, a country about 130 times larger than Panama, only forty-seven bat species are currently recognized. Stunningly, on the tiny island of BCI, seventy-six species of bats belonging to eight different families have been documented.

Most of the bat species on BCI belong to the Neotropical leaf-nosed bat family, or Phyllostomidae. This is a large bat family with over two hundred species that have undergone a remarkable evolutionary radiation, a rapid diversification resulting in a large increase of species. The phyllostomids exceed any other bat family in terms of variation in diet, morphology, social system, roost choice, and many other aspects. The family name refers to a leaflike skin projection that extends from the rostrum, or snout, often in wonderfully contorted shapes, which helps focus their echolocation calls. The fact that there are so many closely related species of leaf-nosed bats with widely different ways of making their livings makes them very attractive for comparative studies, and much of the research on BCI has focused on this family.

Despite the long history of bat research by scientists from across the world, only a subset of the species on BCI has been studied in detail. Among the first to be studied in depth was the Jamaican fruit-eating bat (*Artibeus jamaicensis*), one of the most common leaf-nosed bats on the island. At the time, BCI and the surrounding small islands were formed during the flooding of Lake Gatún, and much of the area was deforested. As a result, wild fig trees, which are pioneer species that play a key role in regenerating forests, were abundant on BCI. To this day, some of these old fig trees can still be seen on the island—yet they are becoming rarer as the forest matures and steals their light. Many figs meant that frugivorous (fruit-eating) bats were abundant, and the study of the fig-eating *A. jamaicensis* was a logical first step. Scientists lovingly call these bats AJs, and early studies of their natural history and foraging behavior laid the foundation for our understanding of their key roles in tropical ecosystems. BCI is

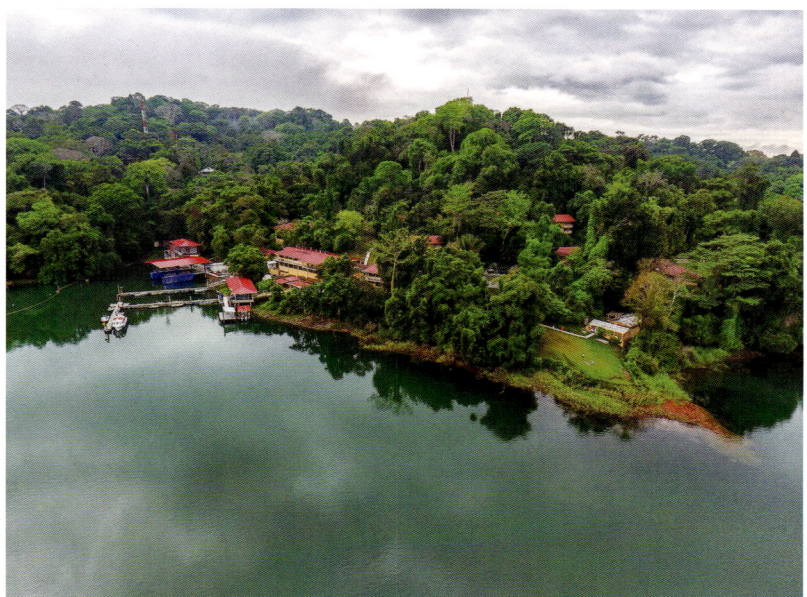

one of the few examples where changes within the tree species community have been closely monitored over decades, and it will be interesting to see how the abundance of certain bat species, such as AJs, respond to decreasing numbers of pioneer fig trees as the forest matures.

As BCI's field station was established a century ago, bats were quick to adapt to the new roosting opportunities. Bat roosts can be found all over the island, with natural roosts in the cavities of giant old trees and under modified leaves in treefall gaps. But now bats have also moved into human constructions, such as the old abandoned lighthouses. For a long time, a colony of small, insect-eating mouse-eared bats (genus *Myotis*) would emerge at dusk each evening from the old steamer *Las Cruces* while she was on the way back to the island from her evening shuttle to the mainland. A colony of the rare long-legged bats (*Macrophyllum macrophyllum*) continues to live in a small half-submerged old boat in the cove near BCI's laboratory buildings. The unique infrastructure of BCI, combined with the closeness of a diverse community of bats, has fostered research that has been ongoing for the past hundred years.

A view of BCI's Laboratory Cove with the red roofs of the STRI buildings emerging through the forest canopy. Boats dock here to transport researchers on and off the island.

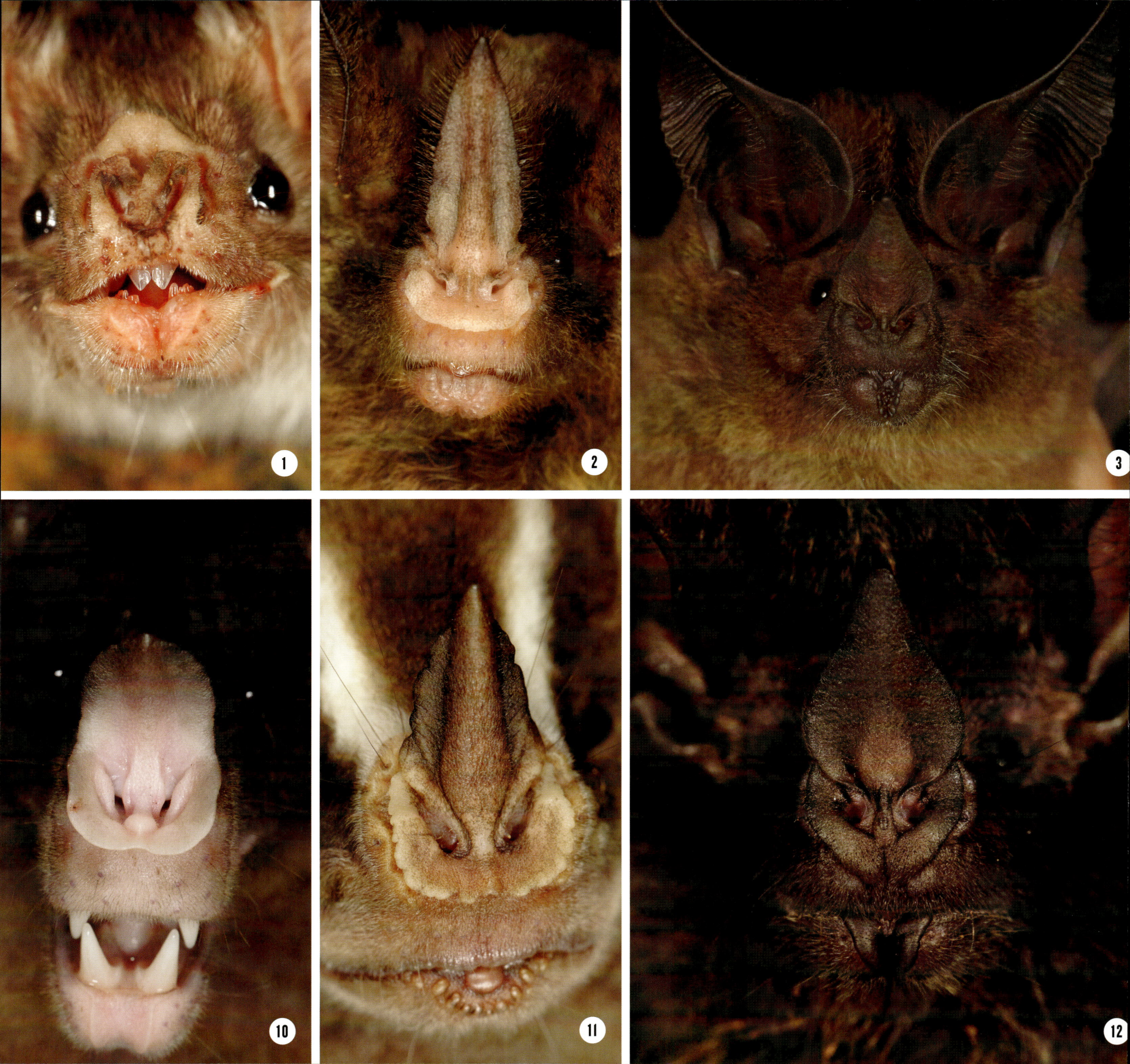

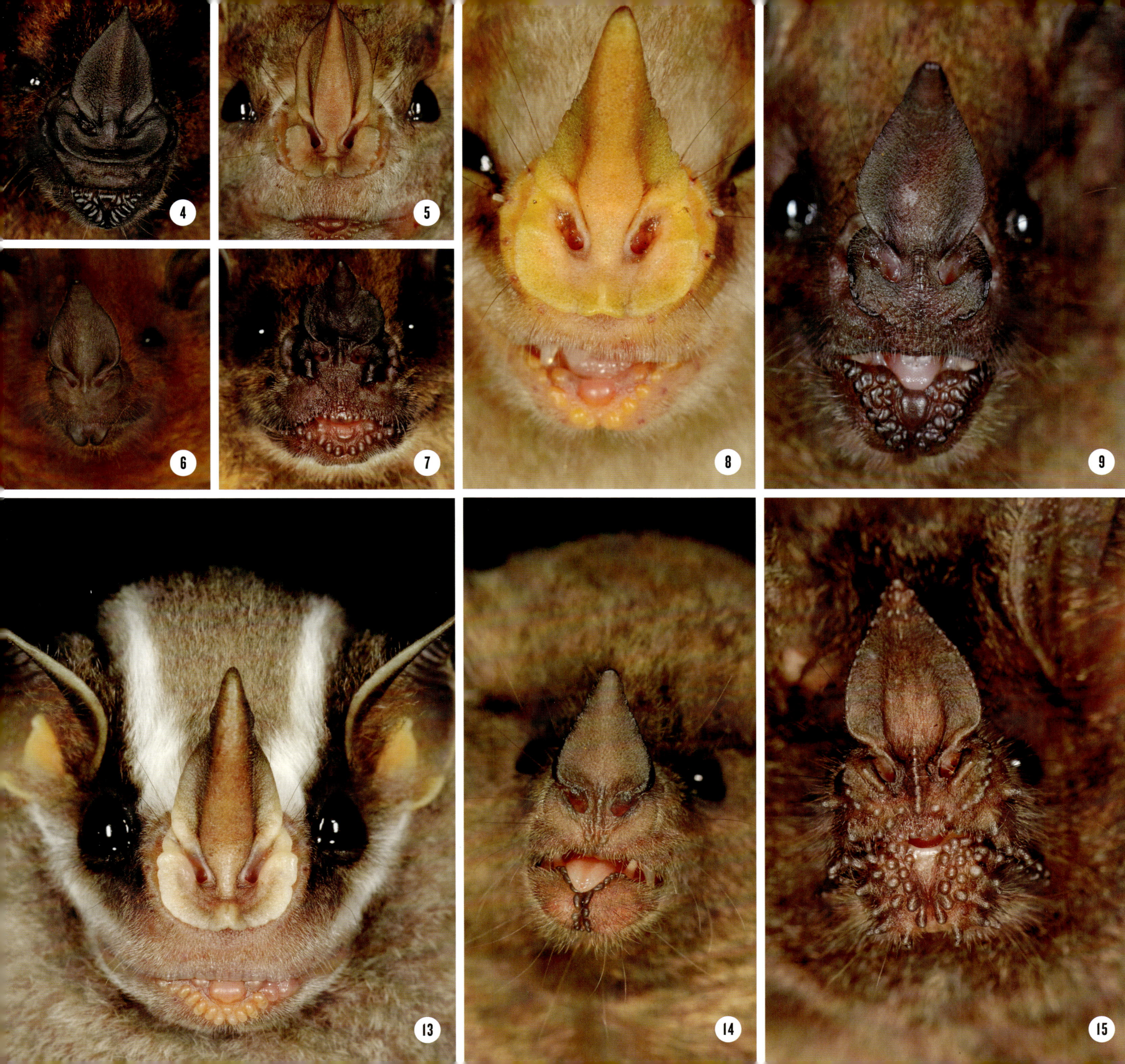

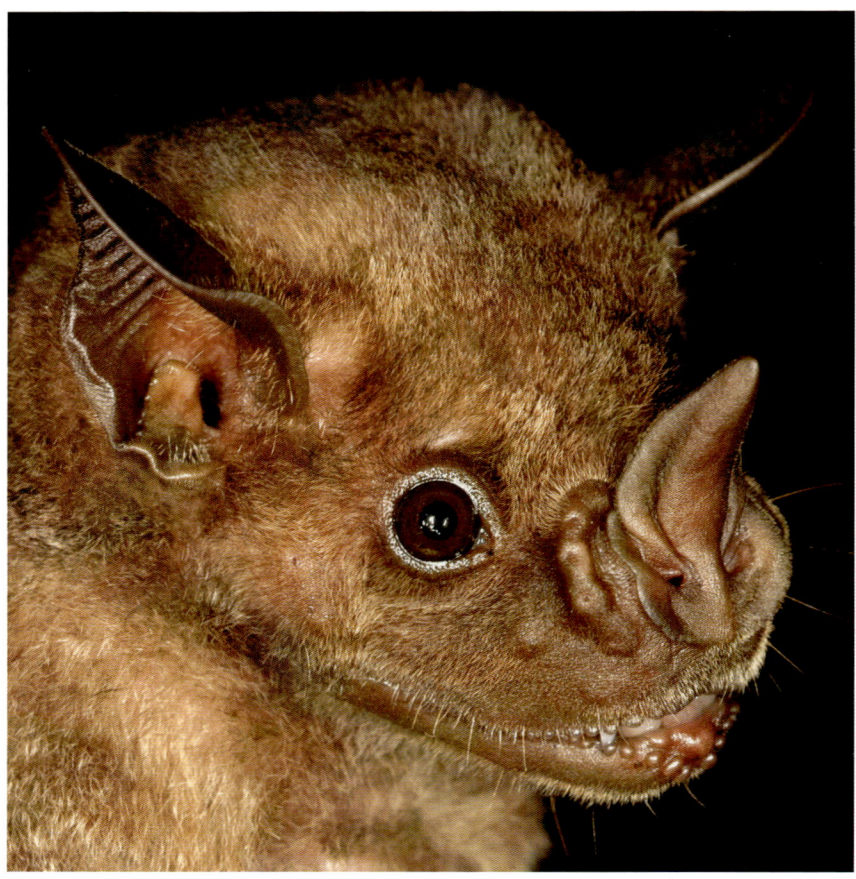

PAGES 16–17 The family of Neotropical leaf-nosed bats (Phyllostomidae) is known for its extraordinary diversity. The spear-shaped nose leaf for which this family is named helps focus the bats' outgoing echolocation calls. Examples shown are 1. Common vampire bat (*Desmodus rotundus*), 2. Striped hairy-nosed bat (*Gardnerycteris crenulata*), 3. Pygmy round-eared bat (*Lophostoma brasiliense*), 4. Greater spear-nosed bat (*Phyllostomus hastatus*), 5. Hairy big-eyed bat (*Chiroderma villosum*), 6. Niceforo's big-eared bat (*Trinycteris nicefori*), 7. Luis' yellow-shouldered bat (*Sturnira luisi*), 8. MacConnell's bat (*Mesophylla macconnelli*), 9. Seba's short-tailed bat (*Carollia perspicillata*), 10. Spectral bat (*Vampyrum spectrum*), 11. Great stripe-faced bat (*Vampyriscus caraccioli*), 12. Common big-eared bat (*Micronycteris microtis*), 13. Striped yellow-eared bat (*Vampyriscus nymphaea*), 14. Commissaris's long-tongued bat (*Glossophaga commissarisi*), 15. Fringe-lipped bat (*Trachops cirrhosus*). ABOVE The Jamaican fruit-eating bat (*Artibeus jamaicensis*) is the most common frugivorous bat on Barro Colorado Island. Photos © Marco Tschapka. RIGHT A pregnant great fruit-eating bat (*Artibeus lituratus*) flies over a fig tree, searching for ripe fruit. Fig trees fruit asynchronously across the forest, creating a near-constant supply of food for the bats—if they are able to locate the fruiting trees.

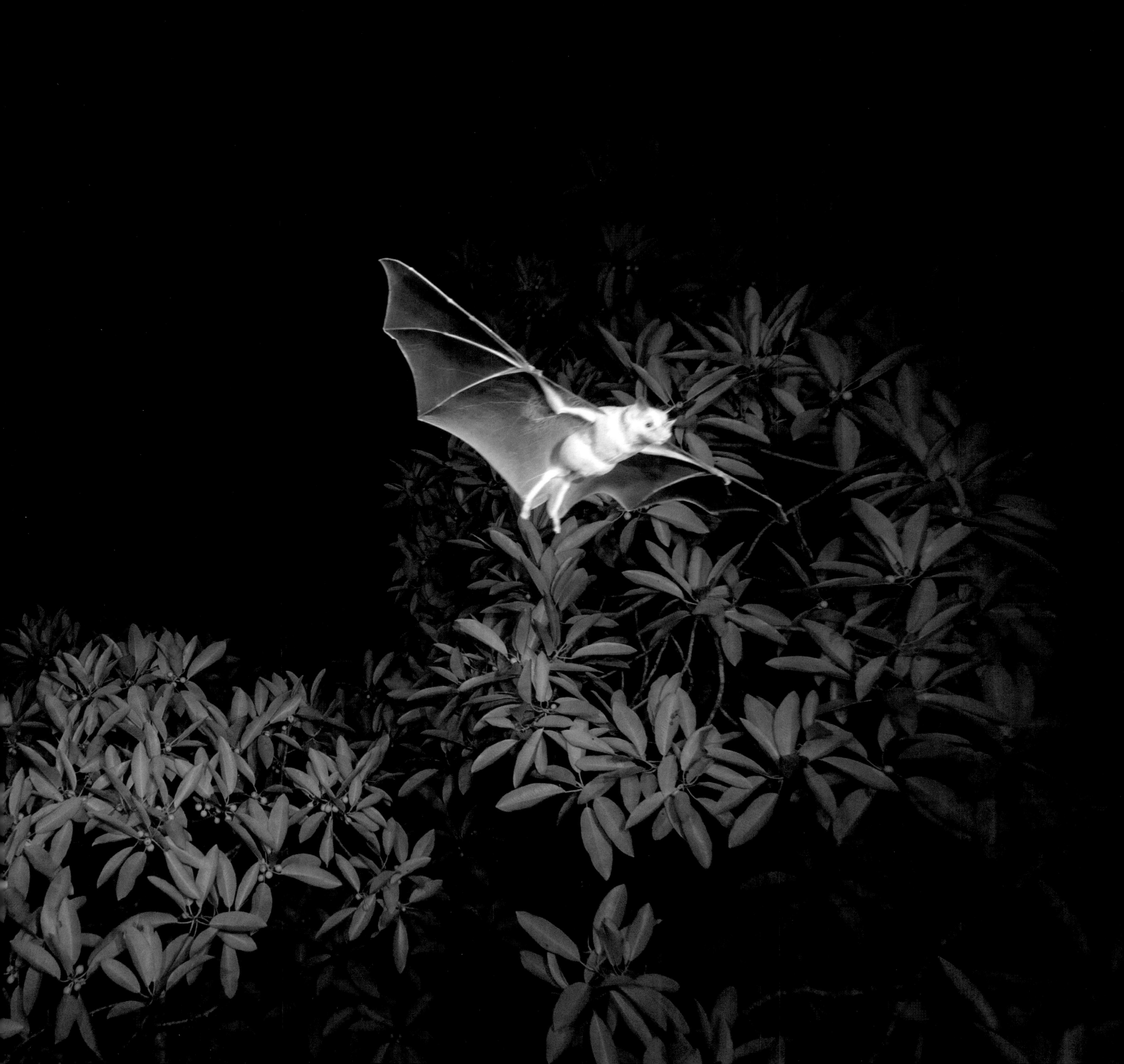

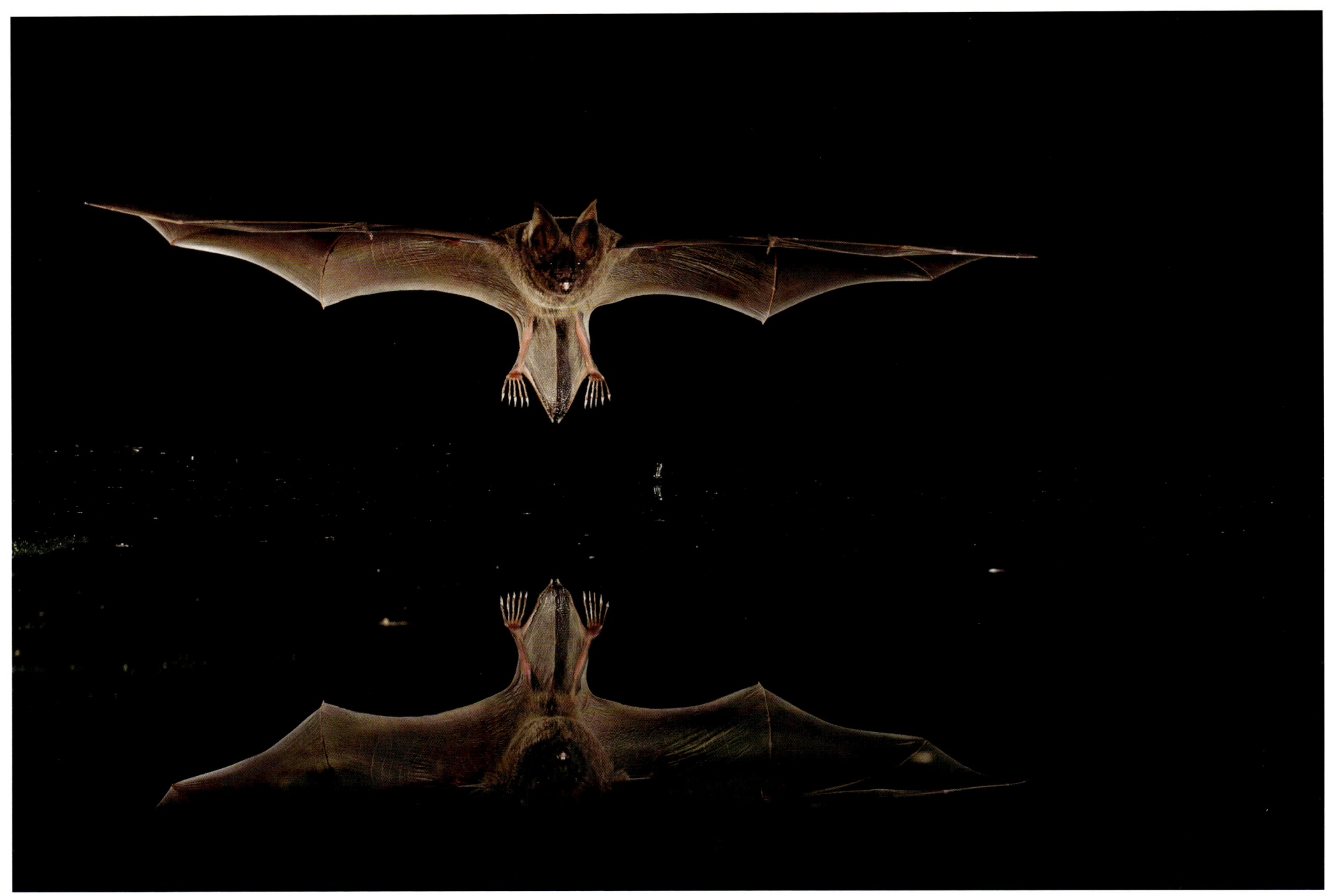

ABOVE Expertly flying above the smooth water surface, a long-legged bat (*Macrophyllum macrophyllum*) descends to capture an insect off the water's surface. OPPOSITE An orange nectar-feeding bat (*Lonchophylla robusta*) visits the flower of a *Quararibea* species. Fruits from these trees (South American sapote or chupa chupa) are commonly cultivated across Central and South America. Photo © Marco Tschapka.

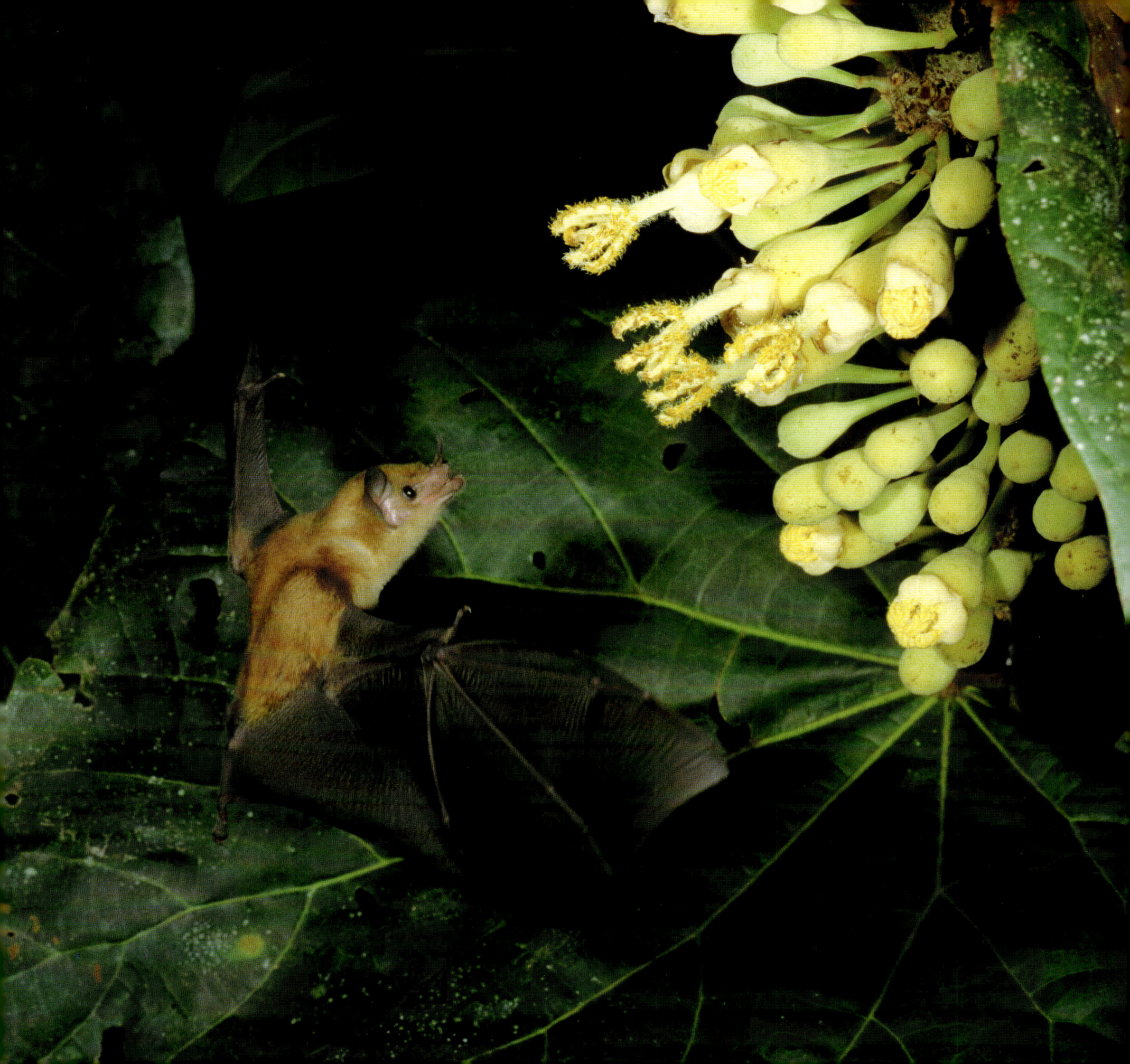

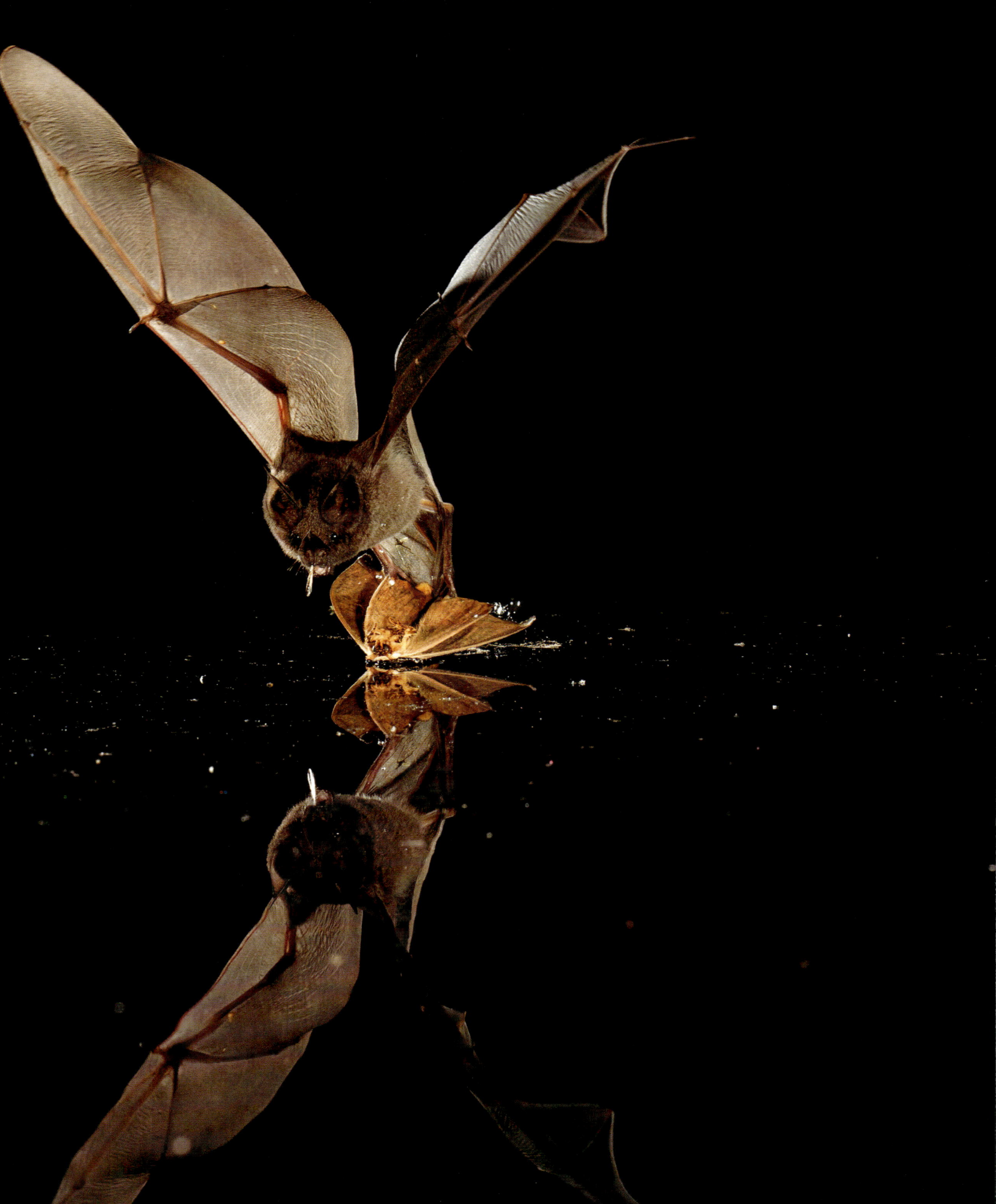

CHAPTER TWO

FLYING HANDS

THE ANATOMY AND EVOLUTION OF MAMMAL FLIGHT

The ecological success of bats, their almost worldwide distribution, and their phylogenetic diversity is in large part thanks to flight. Bats are the only mammals able to truly fly, and the name of their scientific order, Chiroptera, translates to *hand wing*. The oldest nearly complete skeleton of a bat dates from 52.5 million years ago (*Onychonycteris finneyi*) and shows that bat morphology has remained almost unchanged for over 50 million years. These early bats already displayed the enormously elongated fingers of modern bats, thin wings made of skin anchored to the ankles, and the small, sharp teeth of an insect feeder. The specialized hearing structures they needed to echolocate (see chapter 3) would come soon after *O. finneyi*. The only major difference between this ancient bat and its modern counterparts are the claws on the end of each of the four elongated fingers at the edge of the wing membrane, a trait that only modern fruit bats in the Pteropodidae family in Africa, Asia, and Oceania have retained on a few of their fingers.

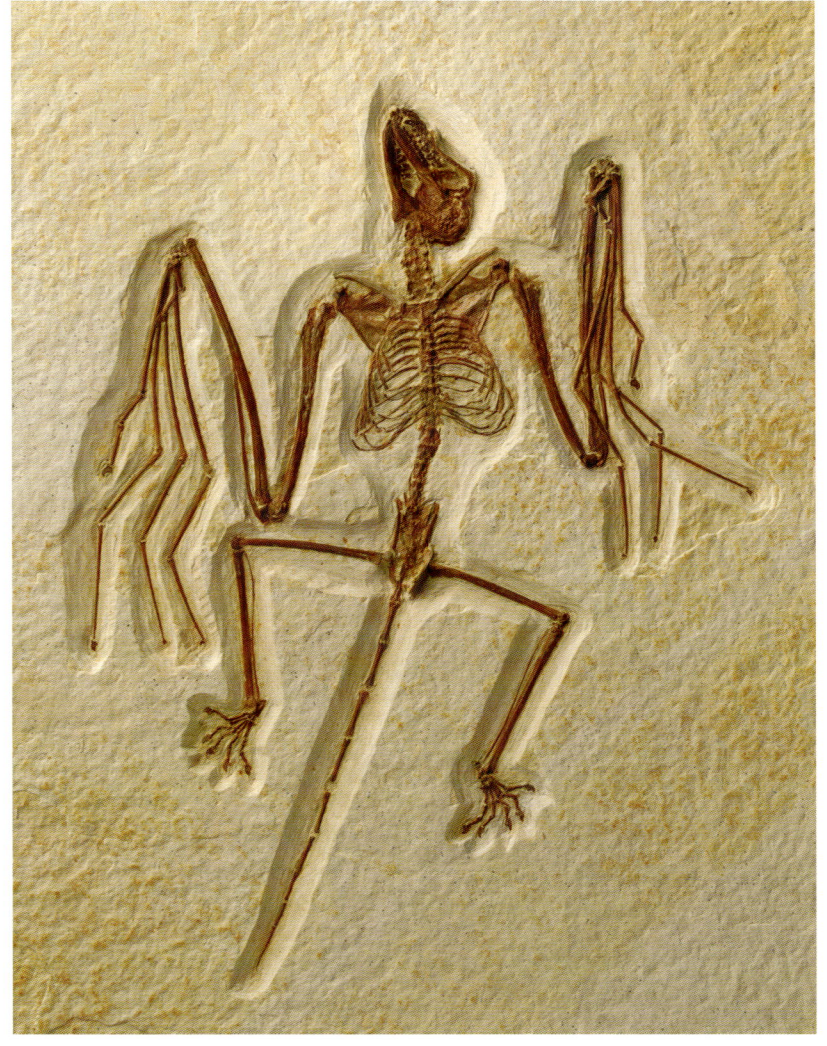

OPPOSITE A long-legged bat (*Macrophyllum macrophyllum*) plucks a moth from the surface of the water as it trawls for food.

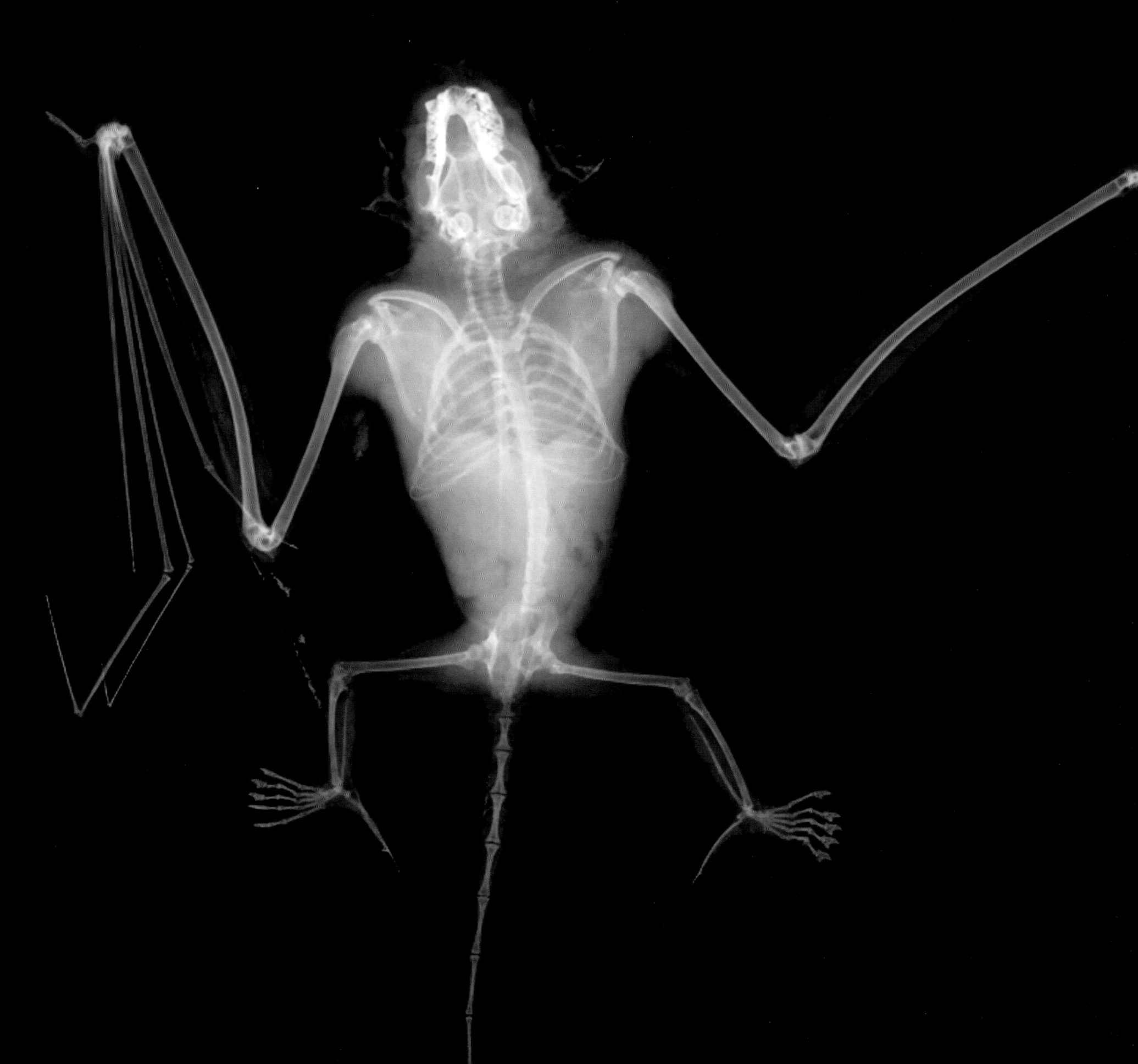

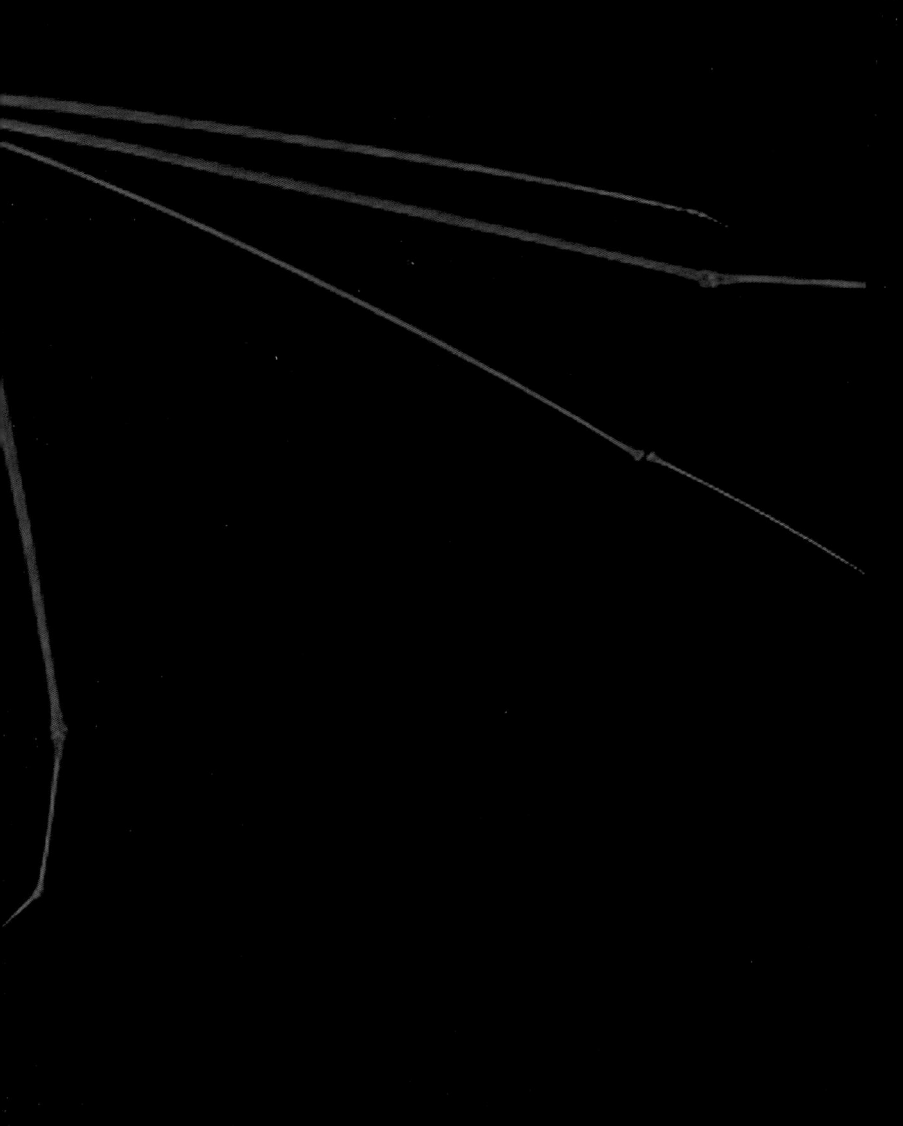

A BASIC OVERVIEW
OF A BAT'S ANATOMY

Like human hands, bat wings consist of five fingers. The thumb is small, has a claw, and projects from the wrist. It can aid with climbing and holding food. The remaining four digits are thin and elongated, connected by a skin membrane called the chiropatagium (or dactylopatagium). A smaller membrane, the propatagium, extends from the shoulder to the thumb and helps shape the wing during maneuvering. The part of the wing between the fifth digit and the body, the plagiopatagium, is anchored to the ankle, and thus creates one continuous flight surface. Many bats have an additional membrane that stretches between the ankles, the uropatagium, or tail interfemoral membrane. Since it is difficult to catch prey in midair with their mouths, many bat species use the uropatagium like a basket to capture insects in flight. Bats flex their knees and do funny little whirls, similar to in-flight sit-ups, to scoop an insect into the uropatagium and deliver it to the mouth. A long spur, the calcar, extends from an ankle bone and helps spread the uropatagium between the tail and hind legs.

Bats can have extraordinarily large ears. The size of a bat's ears is often proportional to how closely it listens to its prey and to its echolocation calls. The largest ears in bats are generally found in species that listen to the sounds of their prey moving or calling. These bat species have a cartilaginous projection, the tragus, at the front of the ear that helps direct sound to the ear. Many bat species also have exaggerated skin folds and protrusions at the top of their nose that help shape echolocation calls into a focused beam when the echolocation calls are emitted through the nostrils. This allows these bat species to have an even more finely resolved echolocation view of their world and potential food items.

PAGE 23 AND LEFT *Onychonycteris finneyi* is one of the oldest and nearly complete bat fossil species recovered, and shows a remarkable similarity to the X-ray image of a modern common noctule (*Nyctalus noctula*) that is found across Europe and western Asia. The long, outstretched fingers of the hand are clear in both images, as are similarities in the rib cage, shoulder blades, and feet. The sturdier finger bones and longer legs of *O. finneyi* indicate it likely was an agile climber in addition to using fluttering flight. Fossil photo courtesy of Nancy Simmons; X-ray image courtesy of M. Teague O'Mara, Dina K.N. Dechmann, and Javier Lázaro.

WINGS AND FLIGHT

For a mammal to make a leap into flight involved a major reorganization of both anatomy and physiology. Elongated fingers provided wings to fly, and to power this costly way of movement, bats developed enormous hearts and lungs that deliver the oxygen needed to fuel their fast metabolisms and powerful flight muscles. In fact, the weight of their heart accounts for over 1 percent of their body mass, compared to approximately 0.7 percent in other terrestrial mammals. Their enhanced lungs and hearts are supplemented by blood that contains a higher-than-average percentage of hemoglobin-carrying red blood cells to transport oxygen. In addition, they can even absorb oxygen from the air directly into the blood vessels in their wings. Bats can control the shape of their wings by moving their fingers and legs, and they can alter the stiffness of their wing membrane with small bundles of muscles located throughout the plagiopatagium, the portion of the wing closest to the body. Bat wings are also covered in tiny hairs that help sense the flow of air, which gives them further control over their outstanding flight acrobatics. A bat's wings reflect its ecology and the environment in which it flies. Some species can fly at speeds over 100 miles (160 kilometers) per hour in search of insects, while others are able to hover in place like a hummingbird to sip nectar from flowers.

When landing, bats face the challenge of fighting gravity to turn upside down, with their feet grasping the cave ceiling, a branch, or other landing surface. Birds can slow down their flight and use gravity to land, but bats need to execute an acrobatic masterpiece before resting. As they approach their chosen roosting spot, they use their wings to shift the direction of their momentum and swing their feet upward like a pendulum to land feetfirst. They do this so well that many bats exert zero force on their landing spot as they land. In fact, there is considerable diversity

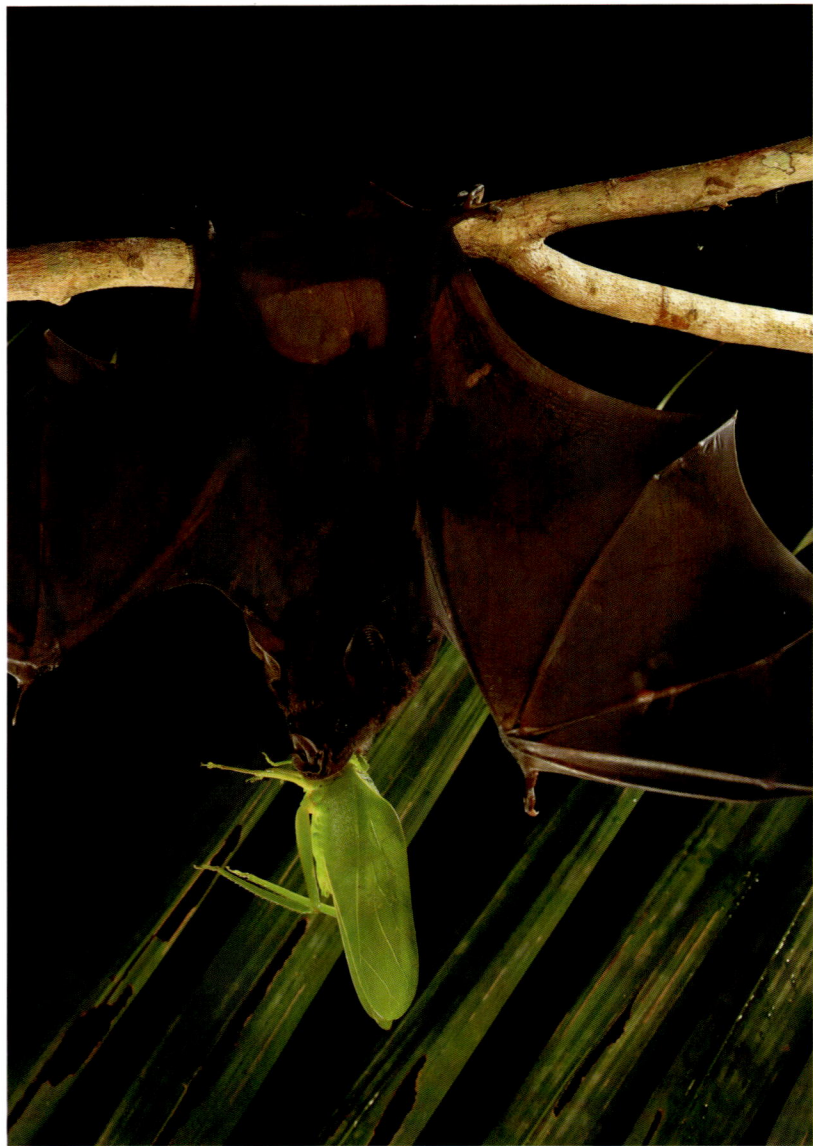

in the ways bats land, depending on the places where they roost. Bats that roost on hard surfaces tend to use these highly acrobatic somersaults, while bats that roost on more flexible leaves and branches will often skip the somersault and crash directly onto their softer landing pad.

ABOVE A greater spear-nosed bat (*Phyllostomus hastatus*) stretches its wings after a successful katydid capture. **OPPOSITE** An overview of bat anatomy. Illustration © Javier Lázaro.

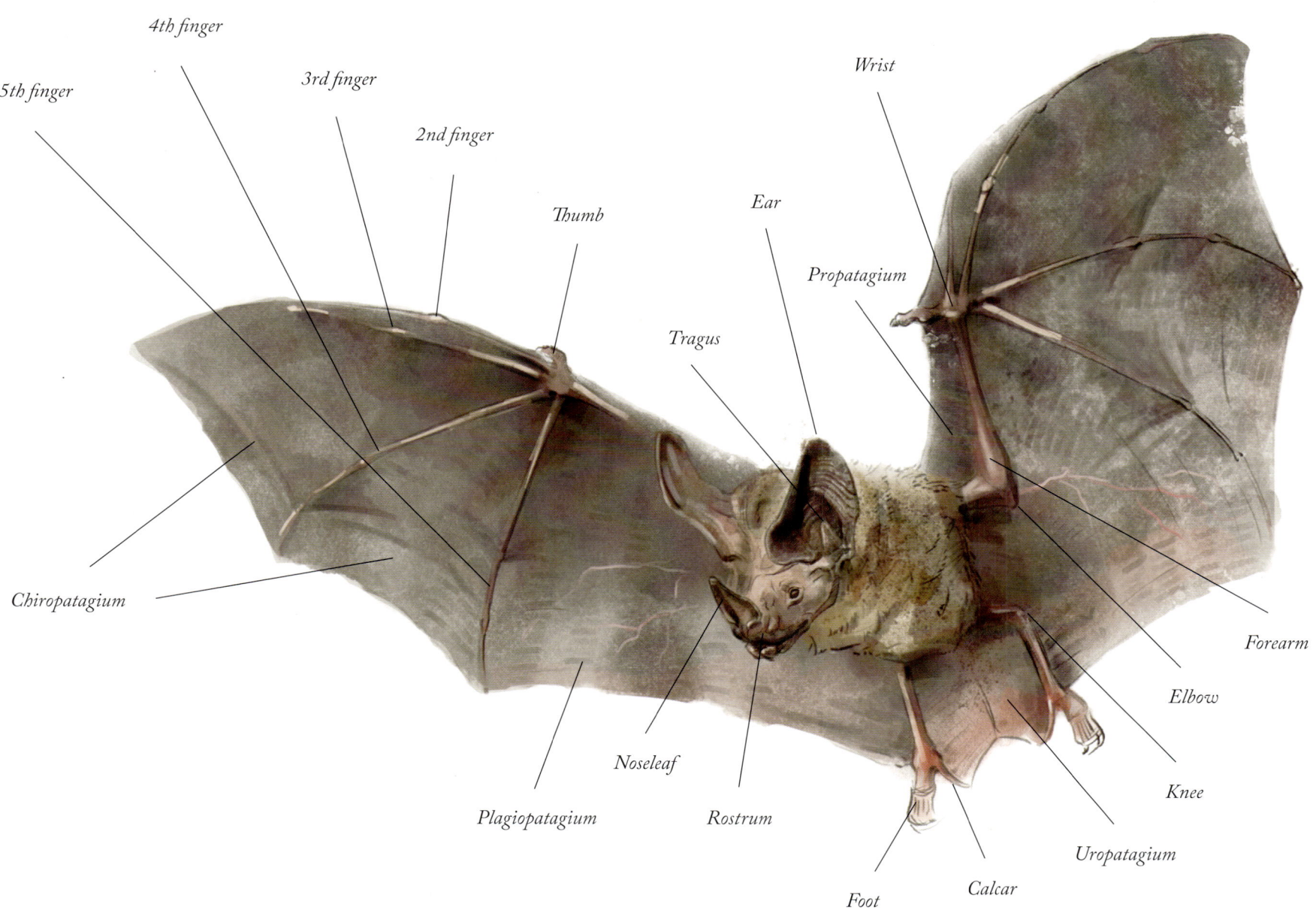

5th finger

4th finger

3rd finger

2nd finger

Thumb

Wrist

Ear

Propatagium

Tragus

Chiropatagium

Forearm

Elbow

Knee

Noseleaf

Plagiopatagium

Rostrum

Uropatagium

Foot

Calcar

HIND LIMBS AND FEET

Resting bats spend most of their time hanging upside down. Specialized anatomy allows them to do this without exerting energy or accruing muscle fatigue, to the degree that even dead bats can be found hanging months after death. To accomplish this, bats have a specialized tendon in their hind limbs that allows their feet to hold onto their roosting surface without effort. In addition, their feet become hooks when relaxed, due to the fact that they have to contract their muscles to release the feet when they take flight. Their hind limbs are rotated 180 degrees—a bat looking down would see its heels instead of its toes. This allows bats to crawl on a flat surface, a tree trunk, for example, without knocking their knees into it. On BCI this feature can easily be observed in the groups of greater sac-winged bats (*Saccopteryx bilineata*) that scamper up and down the vertical walls of the field station's lab buildings, where they roost. Their toes also have long claws that help them find purchase on nearly any surface. Their specialized legs and feet allow pups (baby bats) to anchor to their mother's fur after birth, and, once they are adults, onto cave ceilings, walls, leafy tents, and other roosting structures. To prevent all their blood from rushing to their heads while hanging upside down, bats have a special circulatory system that includes a series of arterial shunts that control blood pressure during flight and when hanging upside down, allowing them to actively distribute blood to where it is needed most.

RIGHT The long feet and claws of greater bulldog bats (*Noctilio leporinus*) allow them to capture fish by raking their feet over the water's surface. CENTER TOP The thin wing membranes of bats are filled with blood vessels, small muscle bundles, and sensory hairs that aid flight. CENTER BOTTOM When outstretched these wings clearly show the bony hand anatomy on which bat flight depends. OPPOSITE By hanging upside down, bats solve a number of challenges associated with flight. Hanging upside down also allows them to use a wide range of roosts that are difficult for predators to access, such as the ceilings of caves or the undersides of leaves. PAGES 30–31 A greater bulldog bat (*Noctilio leporinus*) raking through the water to capture a surfacing fish, showing the reflective, water-repellent surface of its large wings.

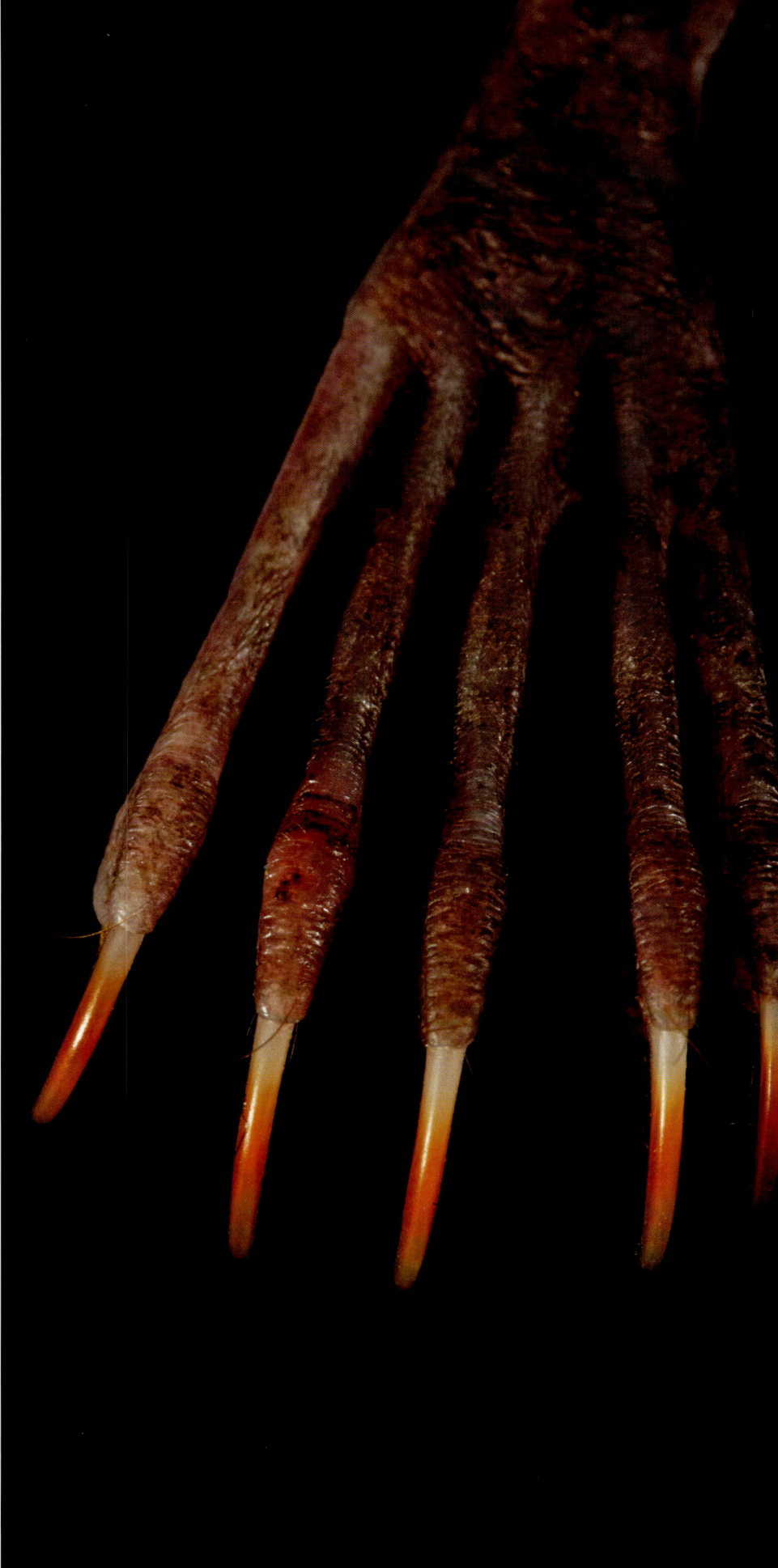

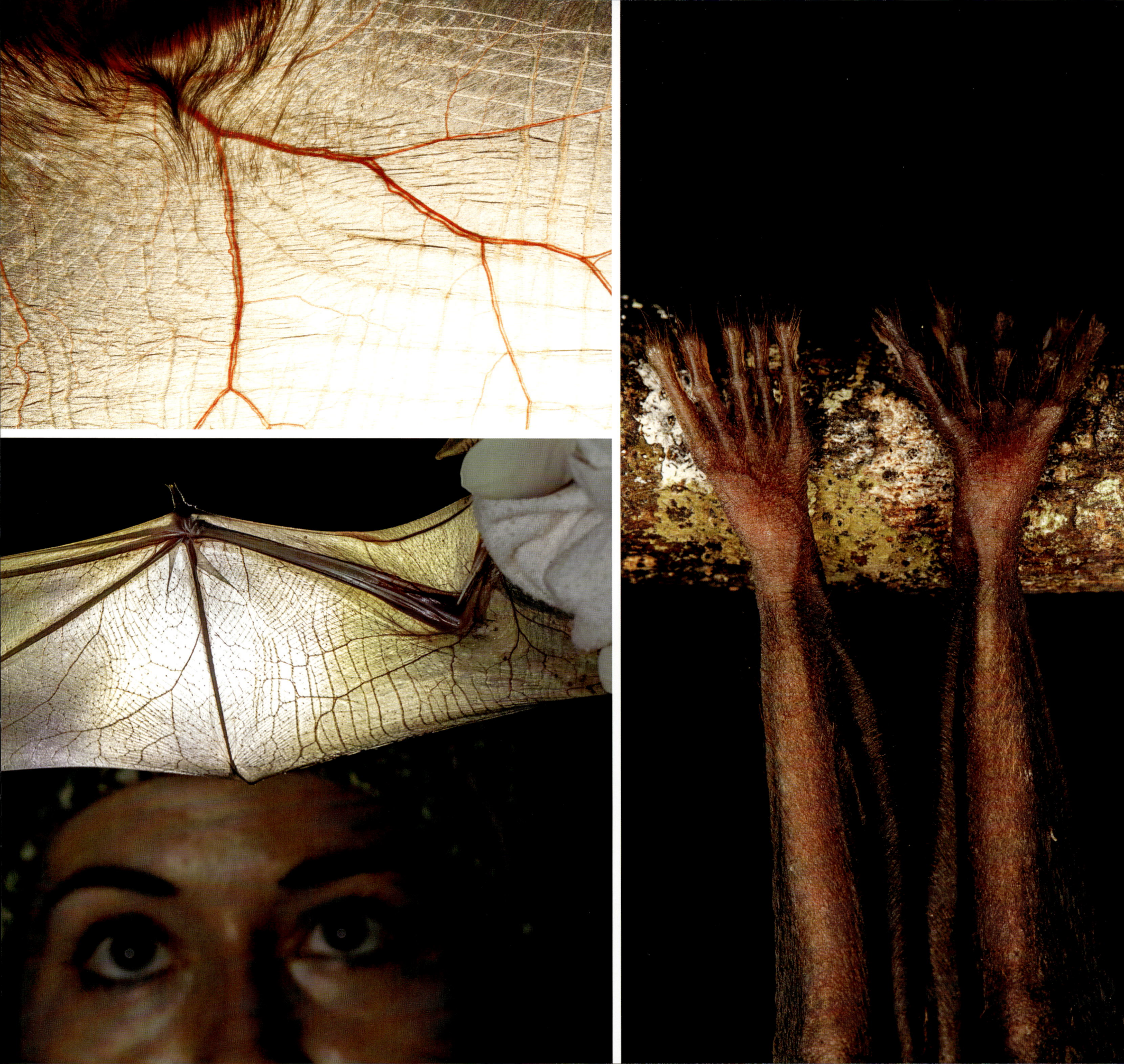

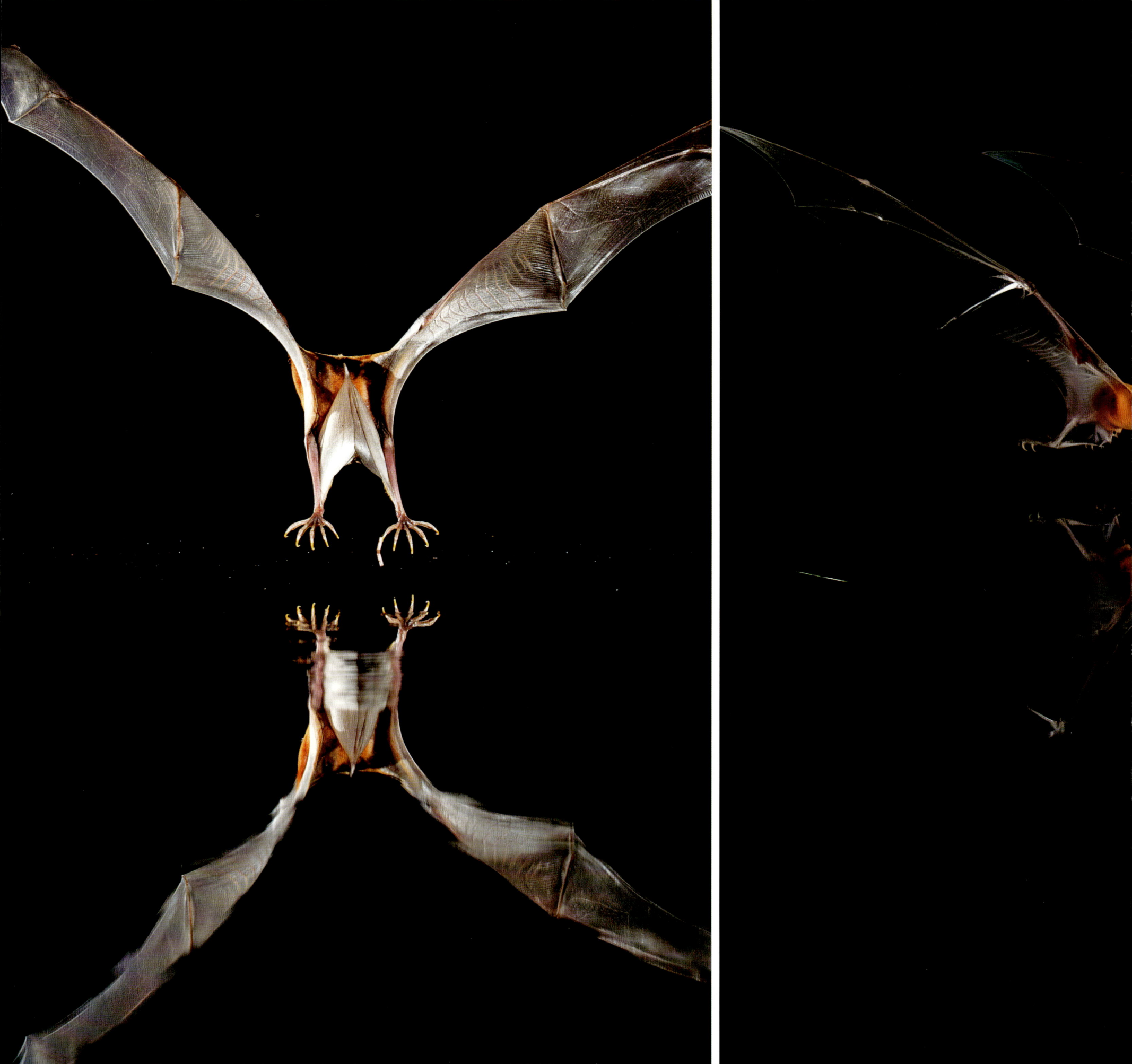

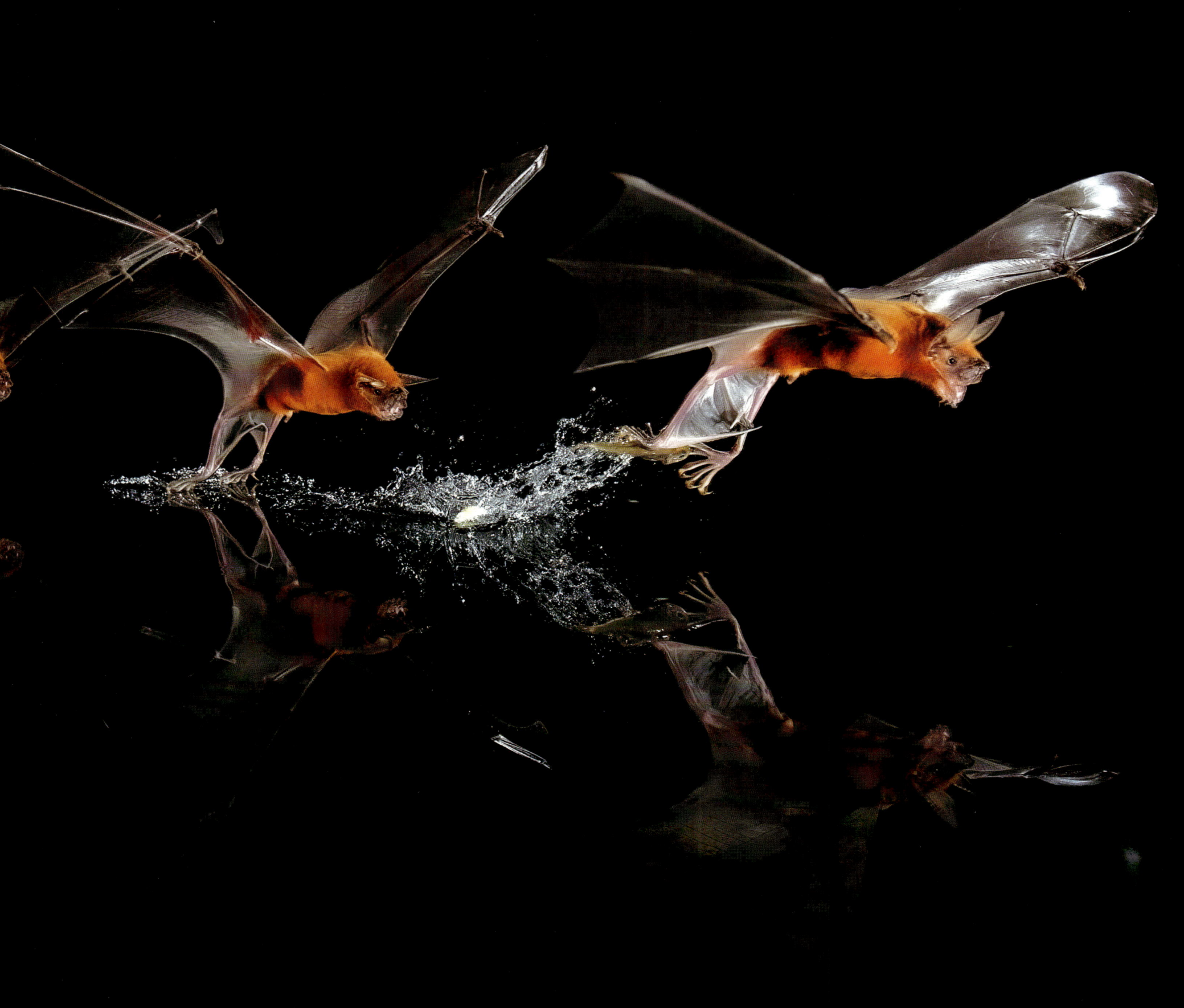

NAVIGATING THE DARK

Echolocation was the adaptation that opened the dark night to bats, allowing them to navigate and feed at a time when birds, the other main group of flying vertebrates, were mostly inactive. Most bats echolocate by shouting at volumes that range from that of a chainsaw to that of a jet engine taking off, and then listening for the echo of that sound to bounce back to their ears. To avoid self-deafening and to be able to hear the much quieter returning echoes, a small muscle in the bat's middle ear contracts when emitting loud echolocation calls, reducing the perceived volume of the call by dampening the movement of the middle ear bones between the ear drum and the cochlea. Their brains then process the echo into an acoustic representation of the world. Bats make these loud, energy-intense sounds in the ultrasound, at frequencies too high for humans to hear, using the same muscles in their larynx that humans use to speak, hum, and sing. The laryngeal muscles of bats are the fastest-moving muscles ever recorded and can sustain vibrations at frequencies up to 95,000 times per second.

Not all bats, however, use supercharged vocalizations to echolocate. Fruit bats and flying foxes in Africa, Asia, and Australia rely mostly on vision to navigate through the dark. Very few of these bat species echolocate, and they use very different forms of echolocation: they create echoes either by clapping their wings together or by producing tongue clicks. Interestingly, humans can use this same tongue click strategy, and some unsighted humans have become so proficient at this form of echolocation that they can ride a bicycle around obstacles. Brain imaging of these human echolocators show that they process the echoes they hear in parts of the brain that are typically reserved for vision.

A fruiting fig tree attracts a wide range of fruit-eating animals. The tree's crown will be stripped of fruit in only a couple of days. Bats typically fly into the fig tree's canopy, select a fruit, and then fly off to another tree to eat in safety, away from the traffic coming in and out of the fig tree. Predators are also attracted to fruiting trees, where they lie in wait for unsuspecting visitors to lower their guard. PAGES 34–37 Multiflash images of greater bulldog bats (*Noctilio leporinus*) over Laboratory Cove at the entrance to the BCI research station allow scientists to understand how these bats use their enormous feet and coordinated echolocation behavior to capture fish.

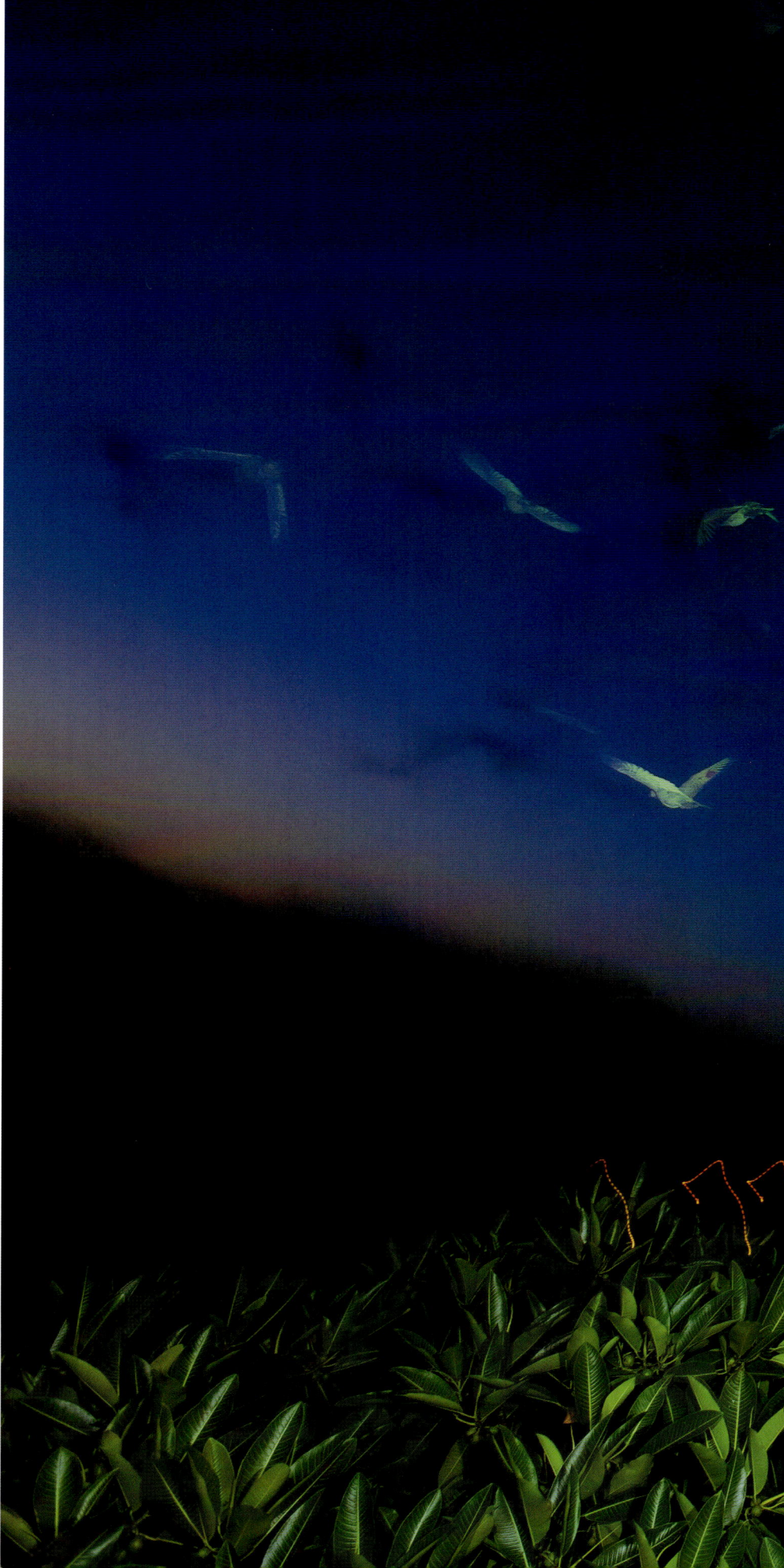

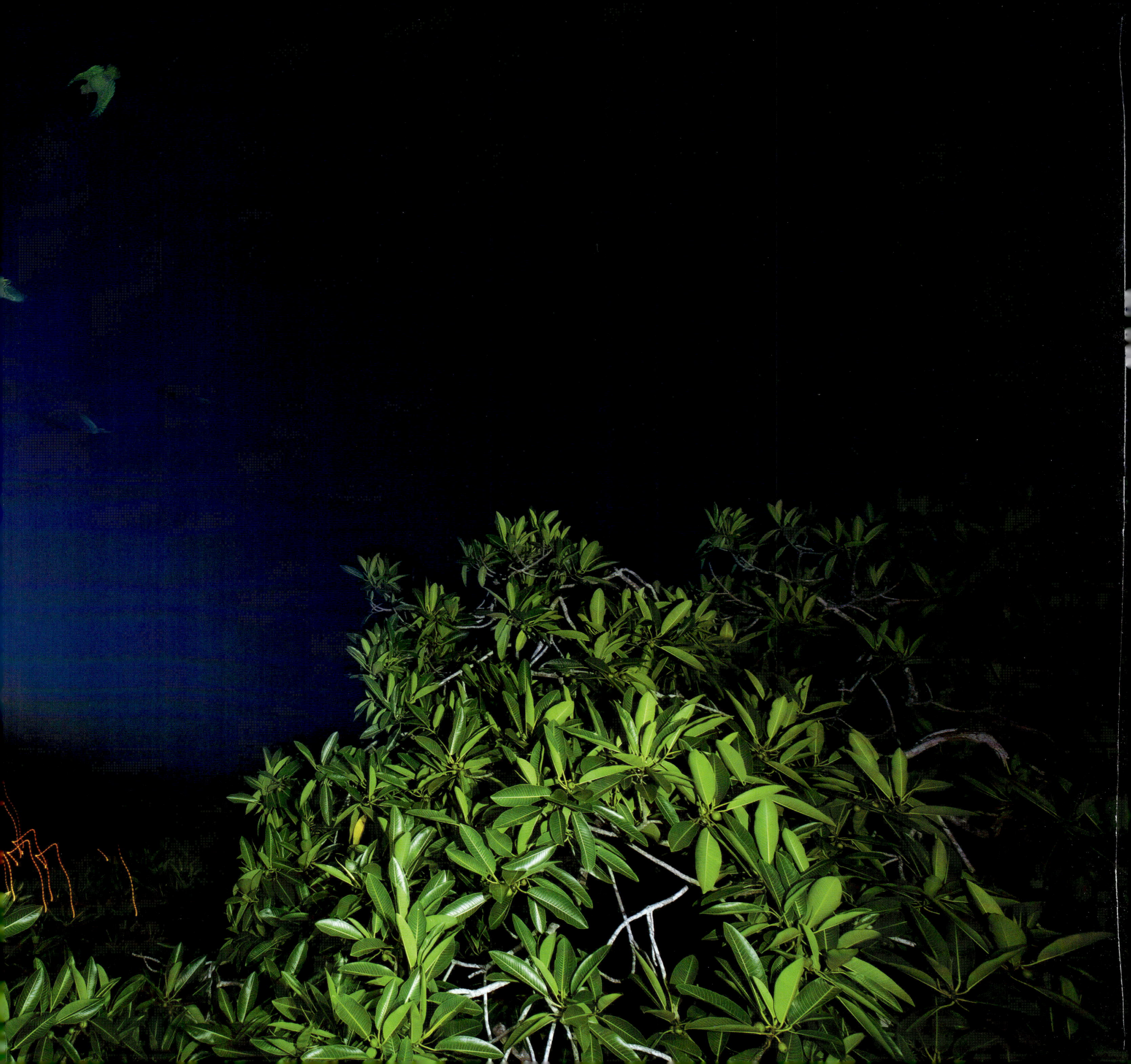

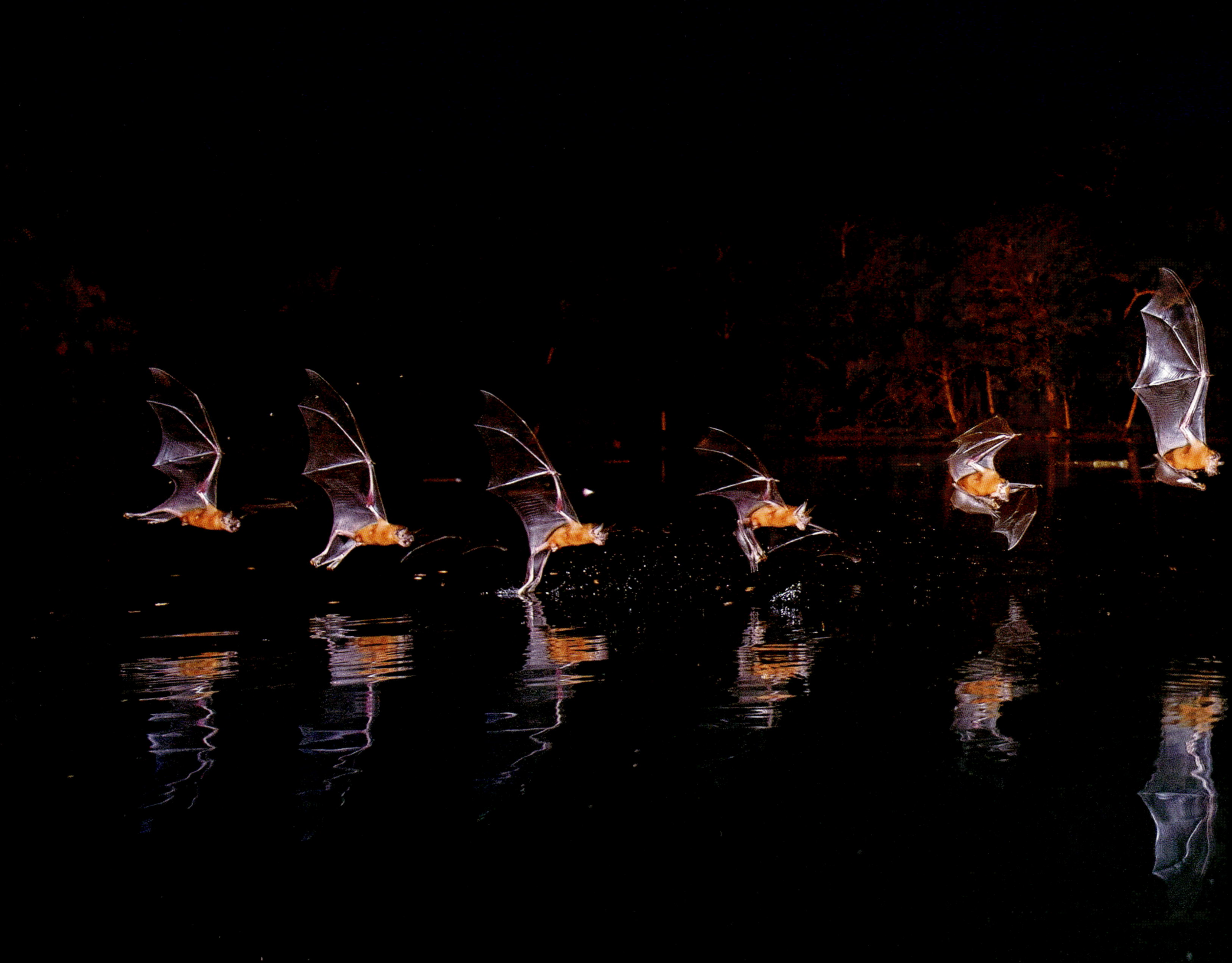

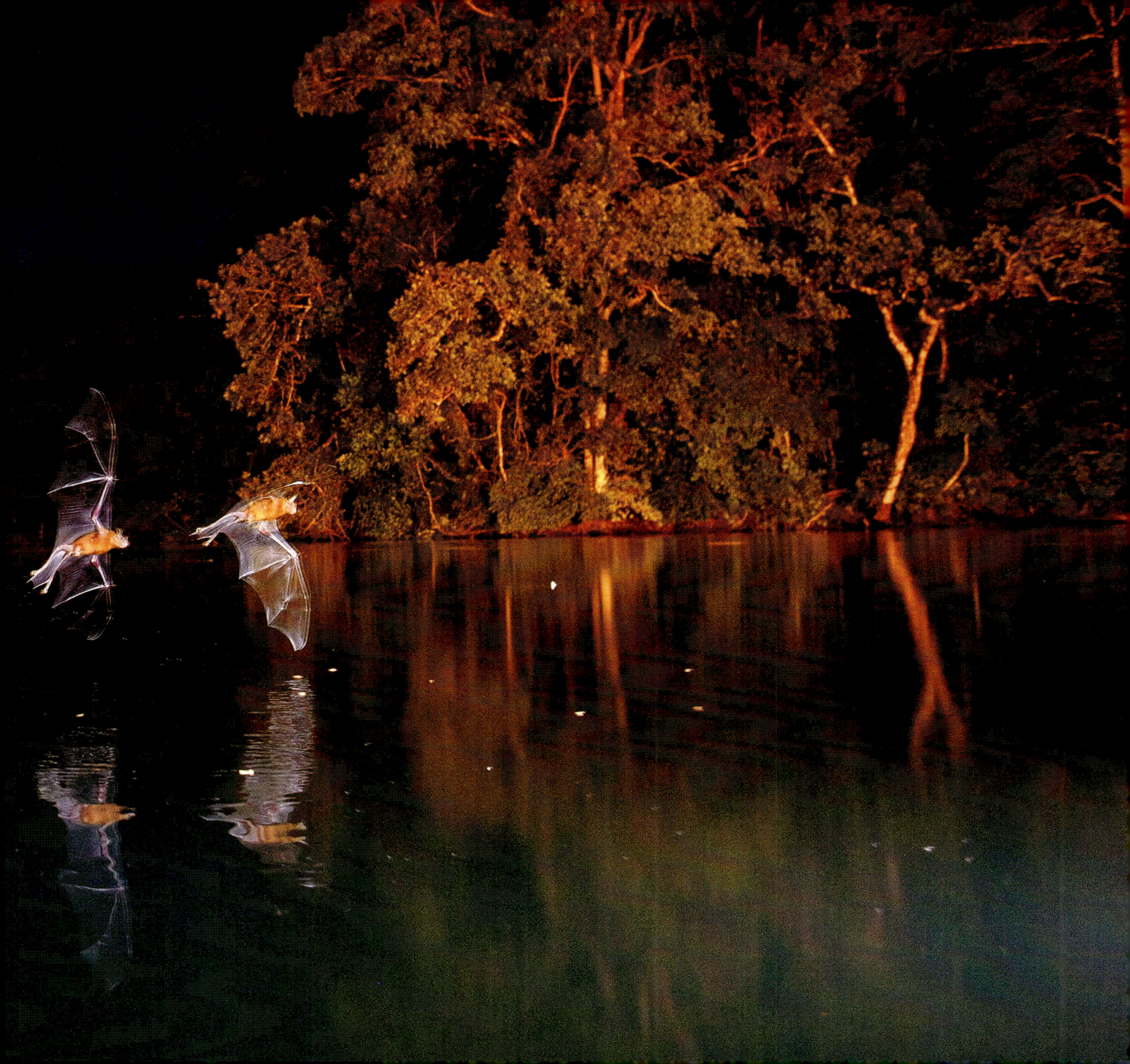

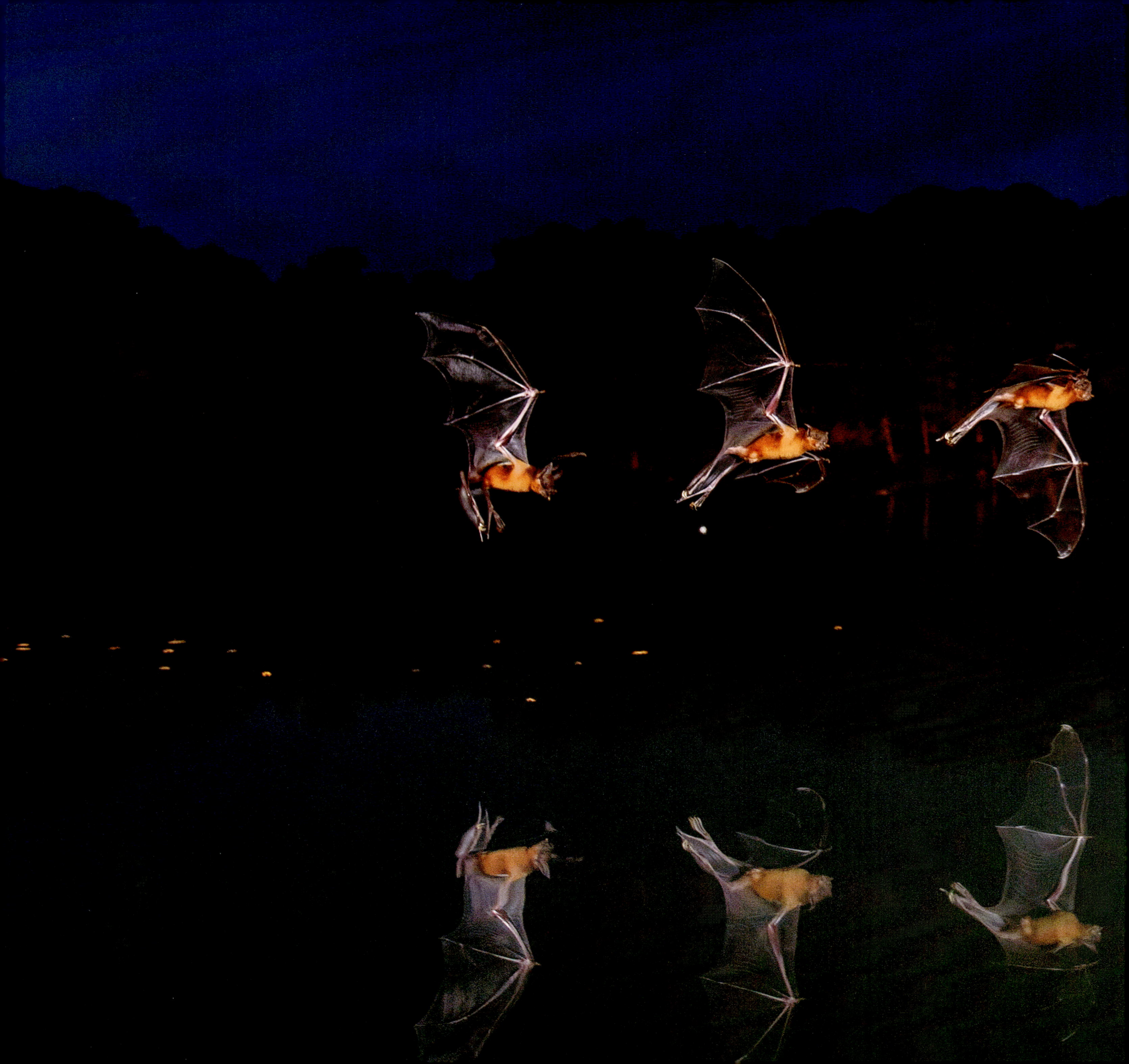

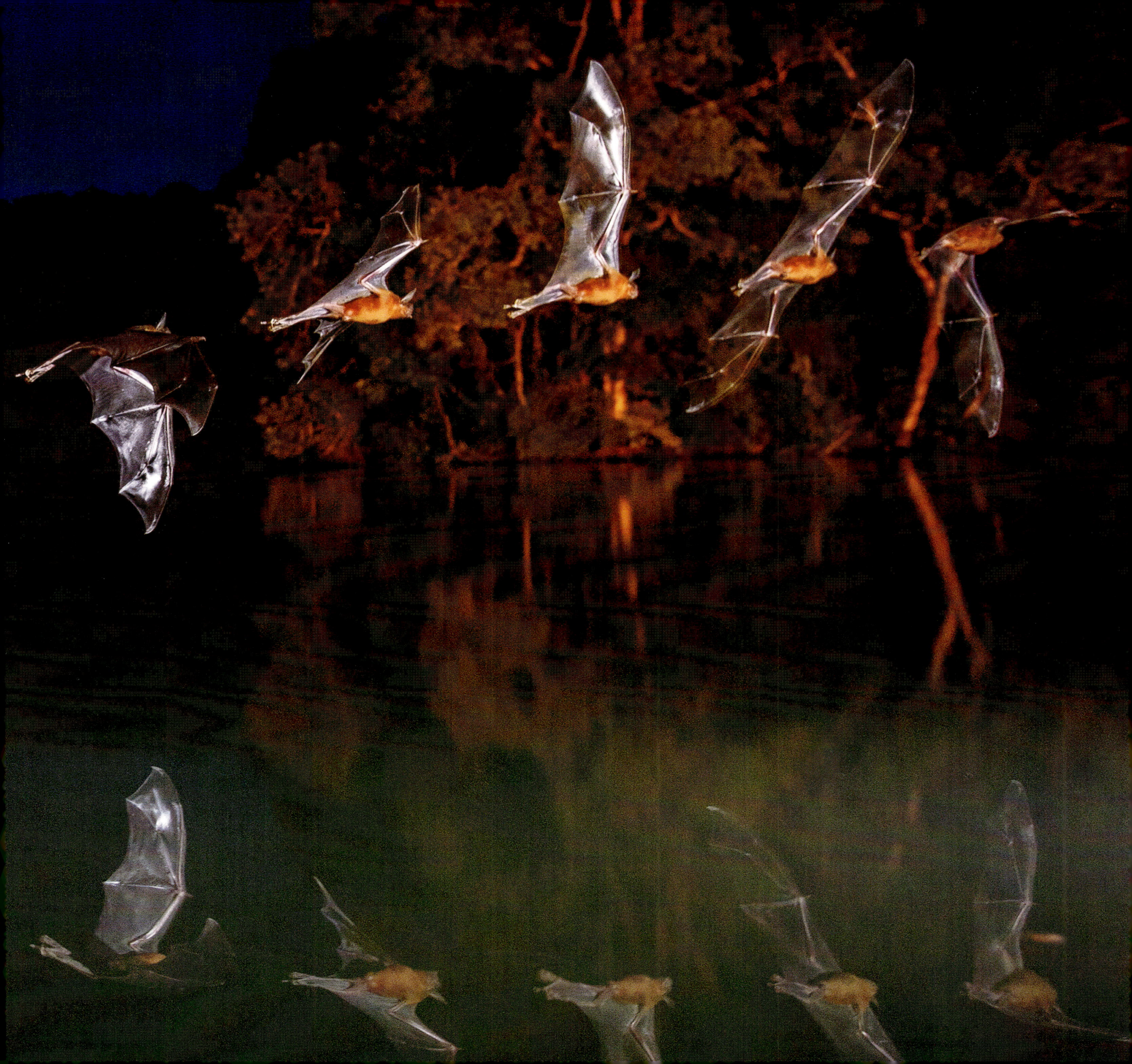

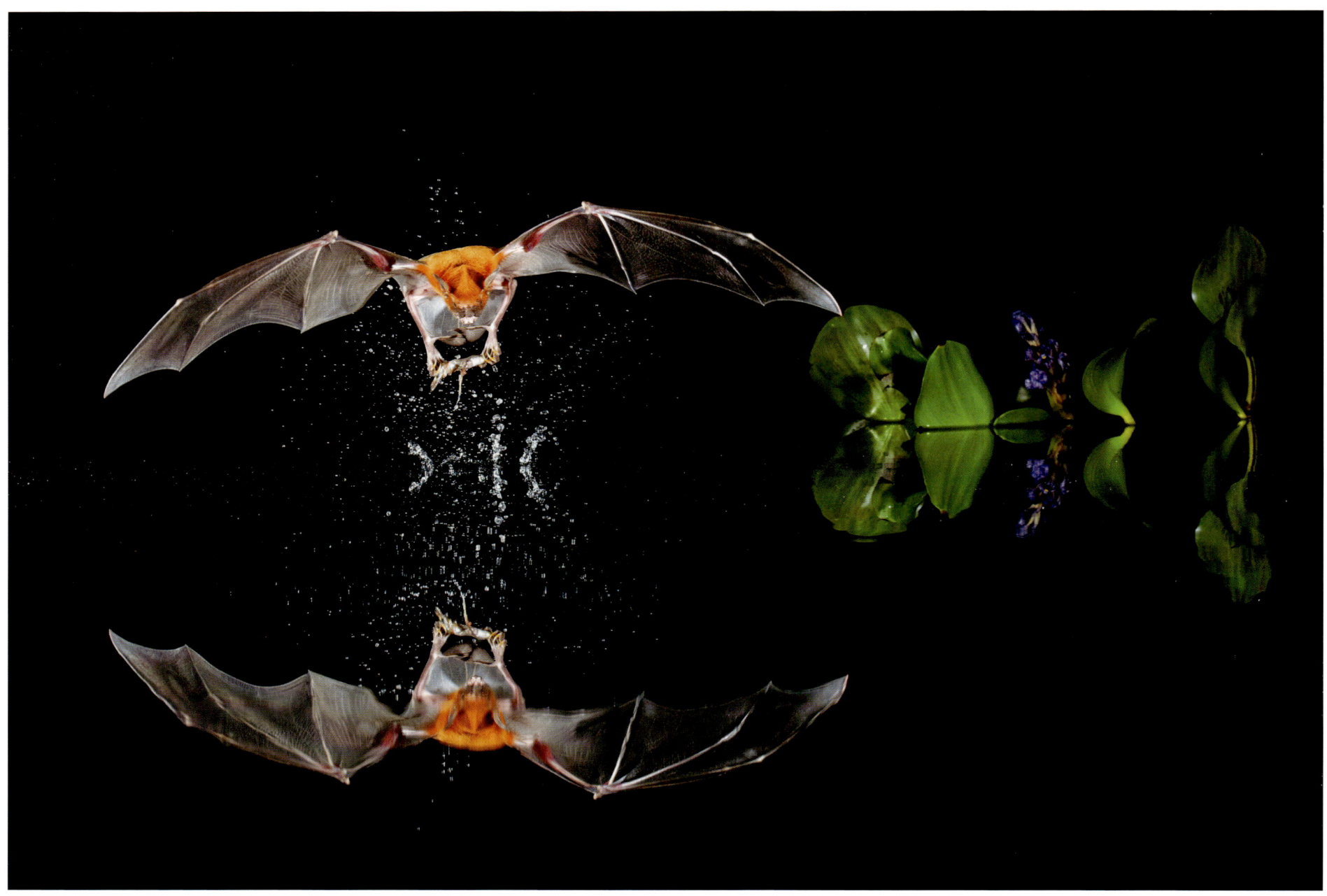

An expert fisher, a greater bulldog bat (*Noctilio leporinus*) flies away with a fish.

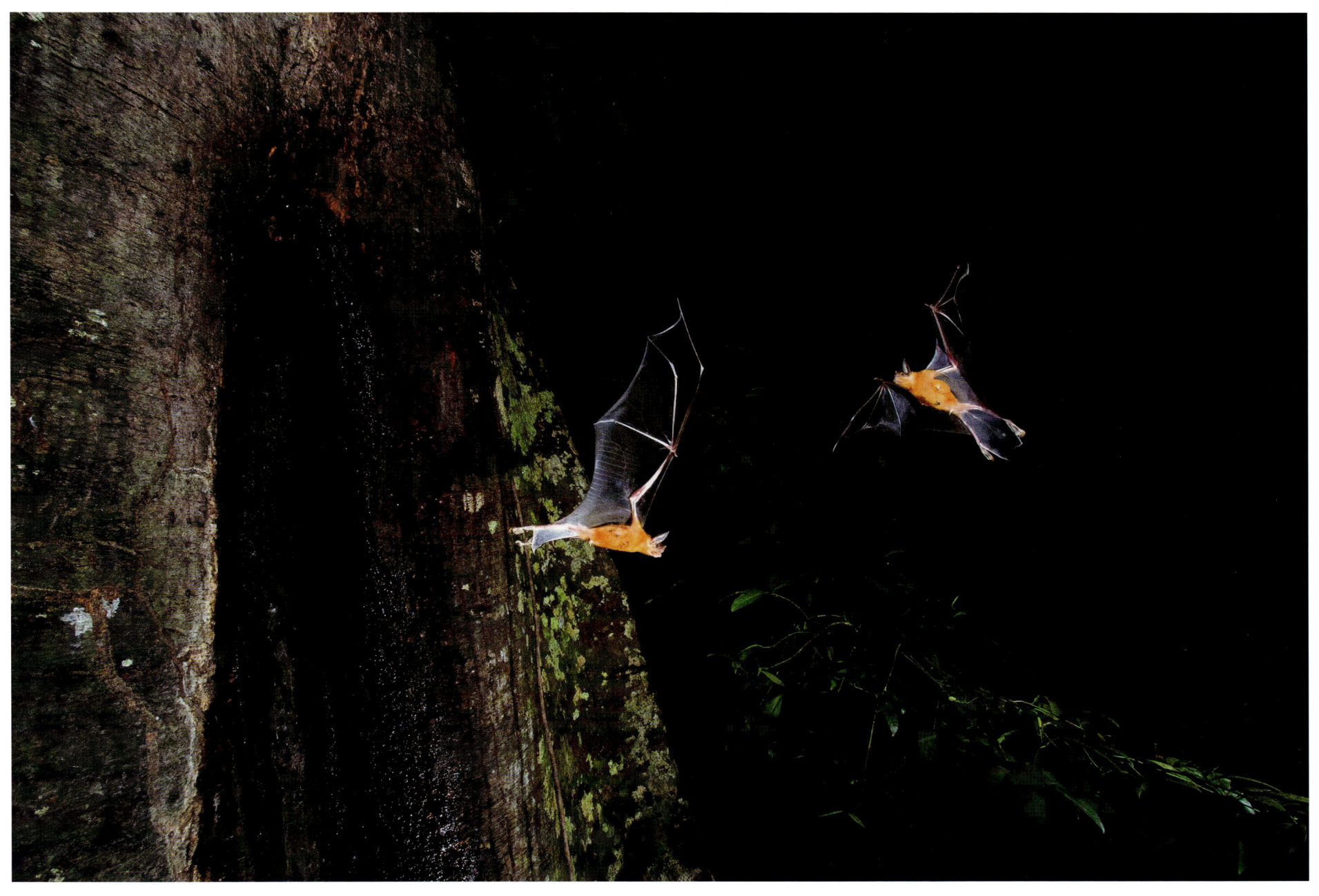

Female greater bulldog bats (*Noctilio leporinus*) leaving their day roost in the cavity of a tree.
PAGES 40–41 A great fruit-eating bat (*Artibeus lituratus*) flying over a fruiting fig tree.

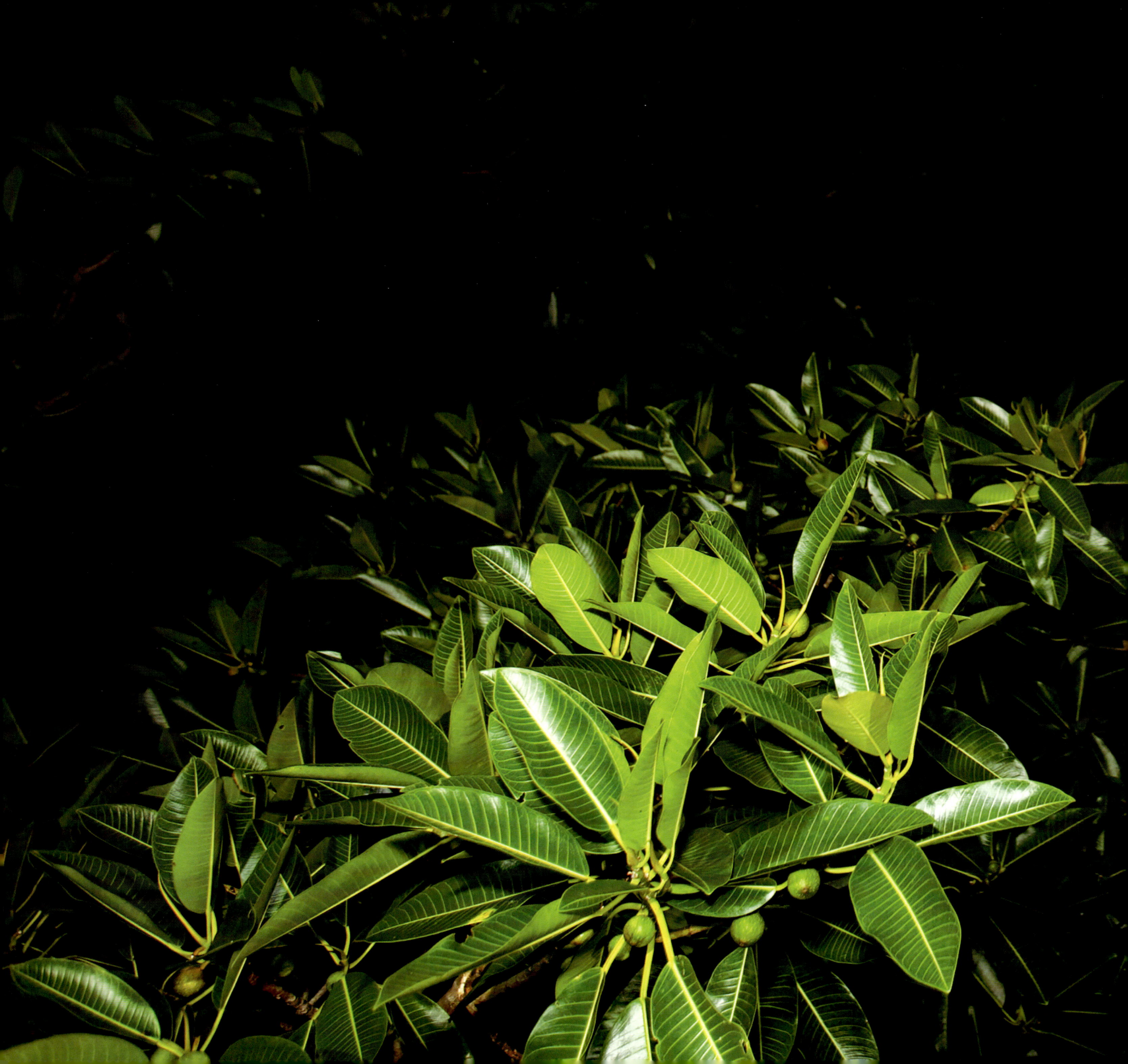

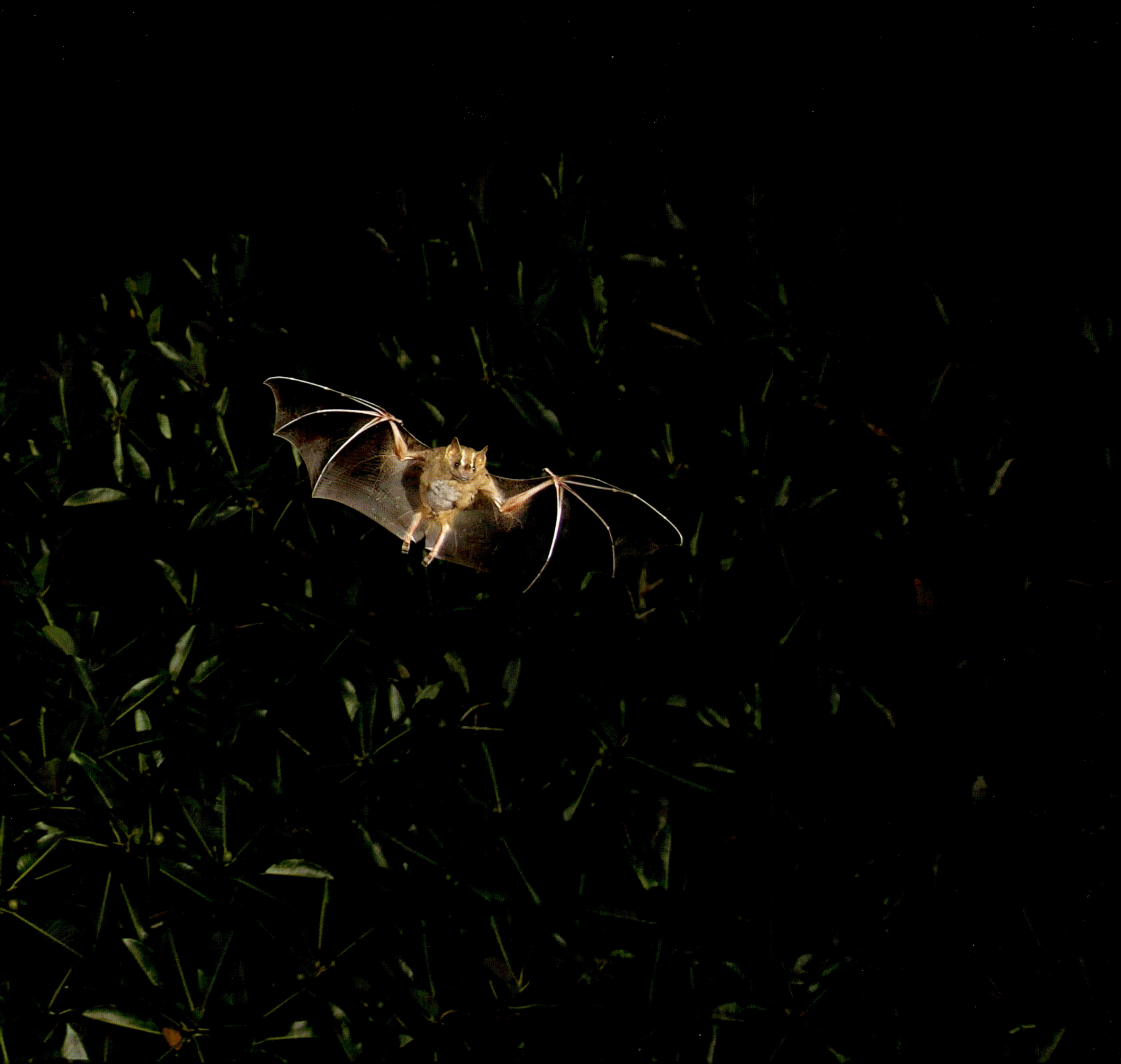

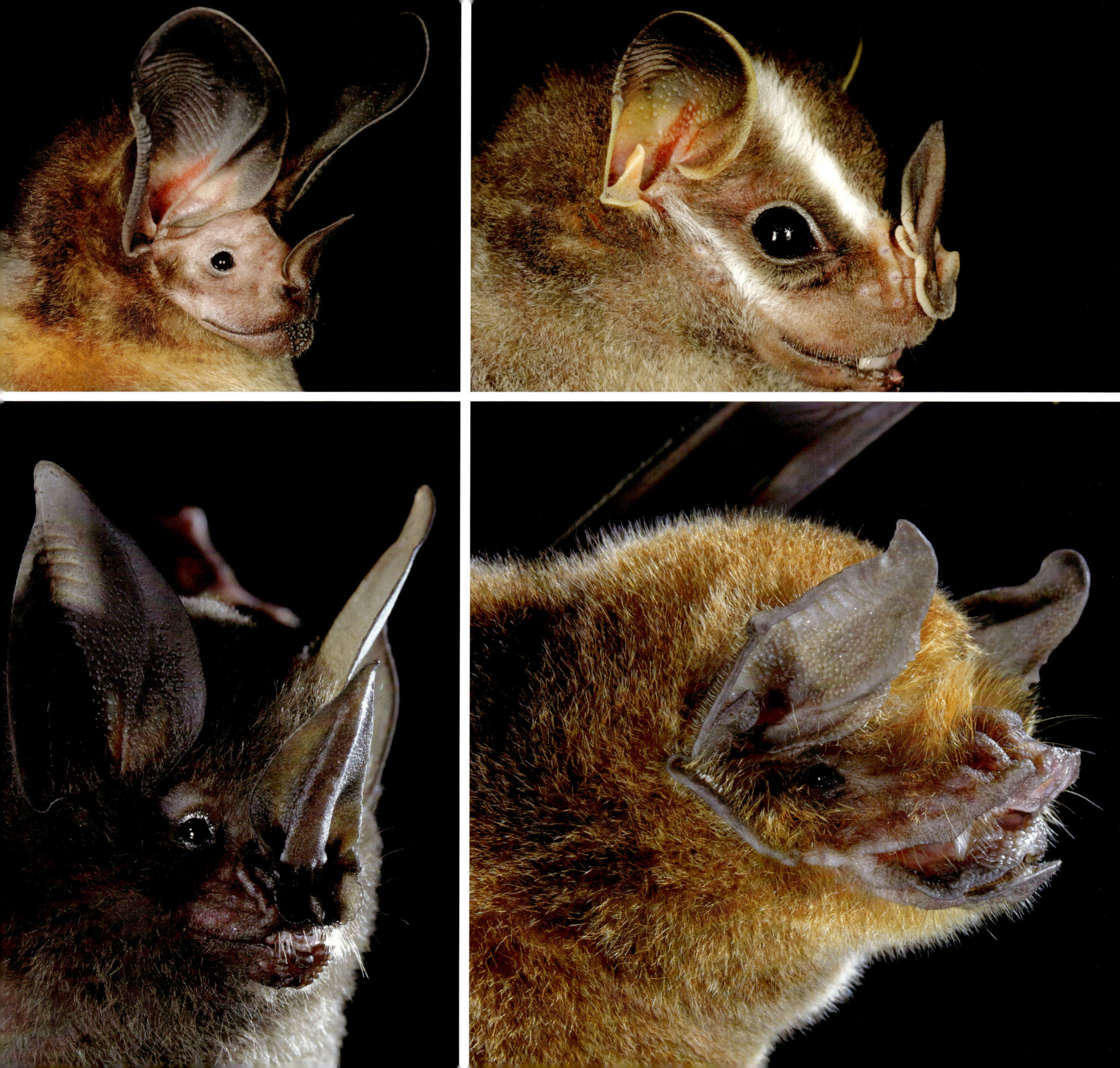

DIVERSITY IN FORM AND FUNCTION

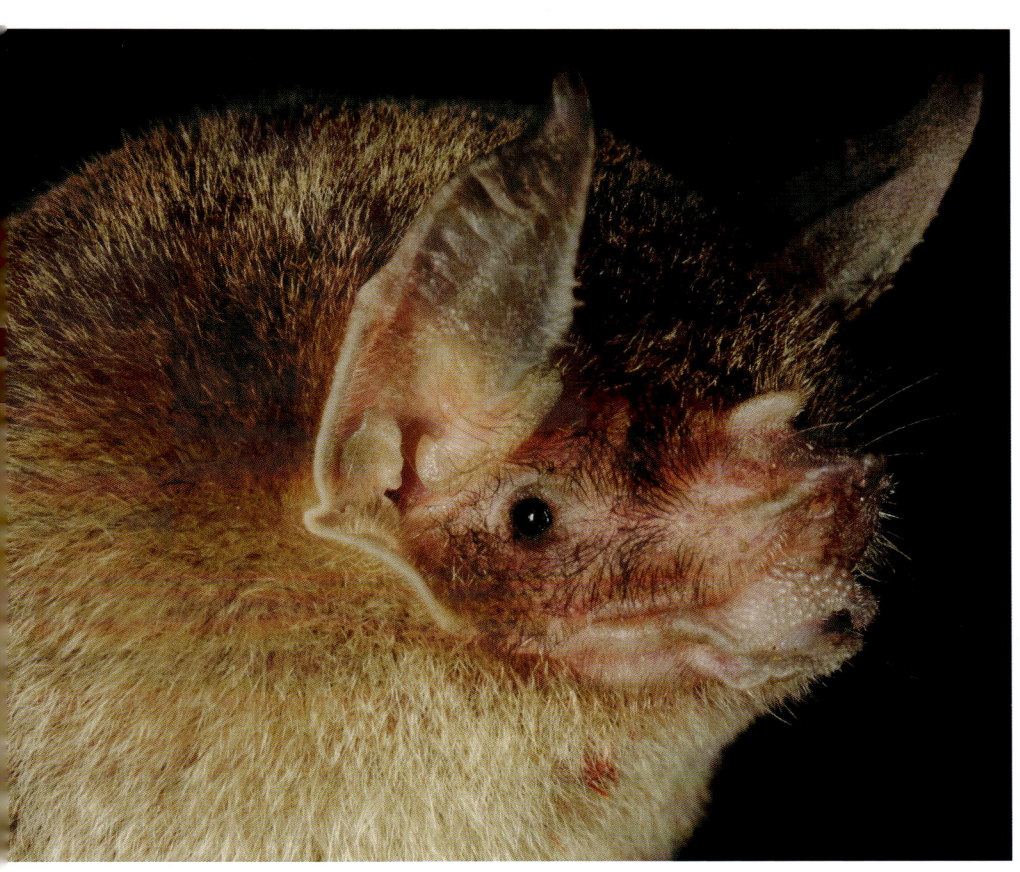

Among the more than 1,450 bat species in the world there is incredible diversity in morphology and behavior that is tightly linked to ecology. The size and shape of the ears, the breadth and length of the wings, and the size of the uropatagium give us a glimpse into how bats live their lives. In addition, their morphological diversity also sheds light on how so many bat species can coexist in relatively small areas. Notably, bats differ in wing and ear shape and these differences reflect their specific foraging ecologies. Bats with long, narrow wings tend to fly high and fast and prefer open spaces, while bats with shorter, broader wings are highly agile and can fly slowly, which is perfect for navigating tight spaces and scanning bushes and trees for food. The largest ears are found in bat species that are sit-and-wait predators that perch and listen for the sounds of their prey—whether that is prey locomotion sounds or mating calls, often of insects or frogs. Since the time of the earliest recognizable fossils of bat species, bats have evolved in ways that allowed ecological expansion and diversity unseen in other mammal groups. This outstanding evolutionary radiation has anchored bats as key members of ecosystems around the world.

BCI hosts a wide diversity of bats of all shapes and sizes. Each has a unique combination of sensory and morphological adaptations, allowing it to access the world in its own specific way. It is this differentiation that allows so many species to coexist in such a small space. **CLOCKWISE FROM TOP LEFT** A few examples include the white-throated round-eared bat (*Lophostoma silvicolum*), Peters' tent-making bat (*Uroderma bilobatum*), the big naked-backed bat (*Pteronotus gymnonotus*), and the long-legged bat (*Macrophyllum macrophyllum*). **ABOVE** Wagner's mustached bat (*Pteronotus personatus*). Top left, right, and above photos © Marco Tschapka. Bottom left and right photos © Adrià López-Baucells.

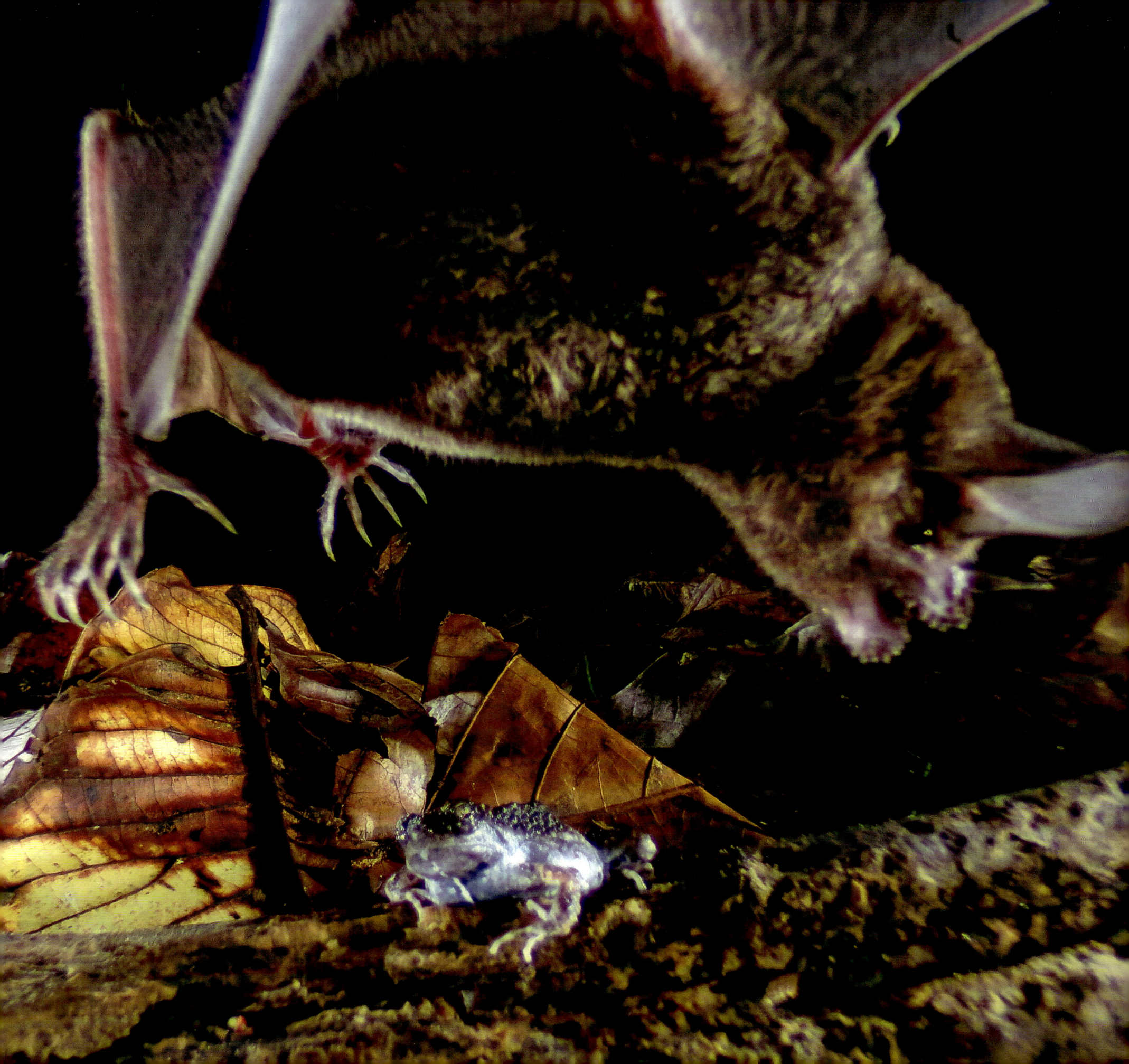

CHAPTER THREE

A BAT'S SENSORY WORLD

NAVIGATING THE NIGHT SKIES

Of all the extraordinary characteristics of bats, the senses bats use to assess their night surroundings are among the most captivating and mysterious. In his 1974 essay, "What Is It Like to Be a Bat?," the philosopher Thomas Nagel famously wrote that it would be impossible ever to understand what it is like to be a bat. Yet decades of research have illuminated some of their many mysteries. Now, we understand their extraordinary sensory worlds better than we ever have before.

ECHOLOCATION

Echolocation was a key innovation that opened a wide set of doors for bats. The roles that birds so successfully fill during the day—catching insects, pollinating flowers, consuming fruits—were largely unoccupied at night. Then, approximately 55 million years ago, bats evolved into these nocturnal niches, which also conveniently helped them to avoid day-active predators. This new echolocation sense soon allowed bats to radiate into a wide diversity of species that could efficiently navigate and find food in the dark. By emitting extremely short bursts of ultrasound and then analyzing the returning echoes, bats can scan their surroundings with such high resolution that even the detection of very small insects in complete darkness becomes possible. Echolocation allowed bats effectively to conquer the night skies.

The diversity we see across bat species is mirrored in the diversity of their echolocation calls. There is not just a single call type; rather, bat species emit a wide variety of echolocation calls that are partly phylogenetically conserved but also adapted to their specific habitat—the edge of a dense forest, for example, the deep understory, or the open spaces above waterways. The duration, frequency, and shape of a particular echolocation call makes it effective in a particular environment. In addition, bats have differentiated their calls by species—to the extent that, for many species, we can identify a passing bat's species simply by recording and analyzing its echolocation calls. Within a species, an individual can flexibly alter its calls to match the immediate demands needed for navigation and foraging. In this way, bat echolocation calls are very different from bird songs. Bird songs signal information to other individuals in a relatively stereotyped way; if the song deviates too much the receiver will not be able to recognize it. In contrast, a bat can flexibly tailor its echolocation calls to the situation at hand. While echolocation calls can also serve communicative functions, the primary receiver of a bat's echolocation call is the bat itself. Therefore, the sender is also the receiver and recognizes its own voice, which allows for considerable flexibility in echolocation call design.

The fringe-lipped bat (*Trachops cirrhosus*), narrowly misses a túngara frog (*Engystomops pustulosus*). This bat eavesdrops on the calls male frogs use to attract their mates. If a frog detects an approaching bat, it falls silent to evade capture by this acoustically orienting predator.

Echolocation not only allows bats to avoid obstacles in the dark, but also to find food. When an echolocation signal hits an insect, for example, a part of this acoustic energy bounces back to the bat, which uses the time difference between emitting the signal to hearing its returning echo to calculate the distance to the insect. Bats have evolved a wide array of strategies to perform this task. Some are fairly simple; others utilize sophisticated calculations to pinpoint prey accurately. Aerial insectivorous bats that hunt flying prey in open space above the rainforest, like Pallas's mastiff bats (*Molossus molossus*), tend to use lower-frequency calls that cover a small range of frequencies. These calls are excellent for covering long distances but provide less information about a target's position and shape. Bats that hunt closer to vegetation mostly use shorter calls with a larger frequency range: these include black myotis (*Myotis nigricans*) and several species from the Emballonuridae family of sac-winged bats, such as the greater sac-winged bat (*Saccopteryx bilineata*) and Thomas's shaggy bat (*Centronycteris centralis*), which fly in treefall gaps, along larger trails, and in other open spaces on BCI. A bat navigating within dense vegetation, such as the Jamaican fruit-eating bat (*Artibeus jamaicensis*) when searching for figs, uses very short calls, since the echo is quickly reflected back to it. With longer calls, the bat would still be emitting its loud outgoing call when its quieter echo returned, and thus would not be able to detect it.

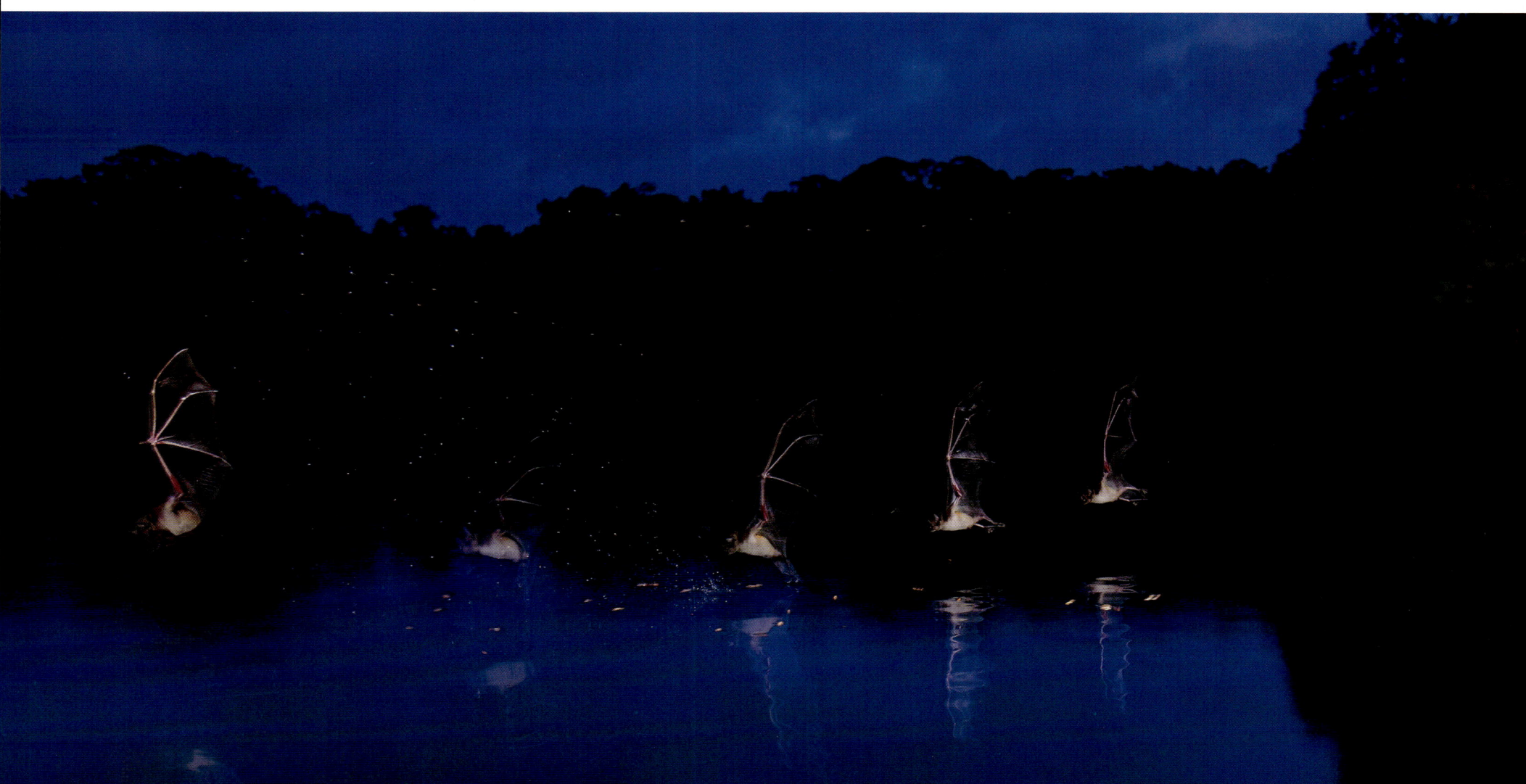

Once an aerial insectivorous bat hunting in open space gets close to its flying prey, it needs more detailed information on the target's location and movement, so it decreases the intervals between its echolocation calls and emits very short calls in rapid succession to obtain high-resolution information. This rapid succession of echolocation calls associated with this terminal phase of the hunting approach is termed a "feeding buzz." In order to hear the echoes of tiny targets, such as a small flying insect, the calls of aerial insectivores need to be incredibly loud. Bats are among the loudest of all terrestrial animals and can produce sounds of up to 140 dB, well above the human pain threshold. Fortunately for us, bats primarily echolocate in the ultrasound range, which

humans cannot perceive—otherwise their calls would bombard our night soundscapes. Occasionally, we can hear chirping sounds from bats darting around at night, but they are generally the lower frequency social calls of bats communicating with each other.

In contrast to the aerial insectivores, bats foraging on non-flying prey in dense environments such as forest understory use shorter, higher-frequency calls that span a broad range of frequencies. Most Neotropical leaf-nosed bats fall into this category, such as the stripe-headed round-eared bat (*Tonatia saurophila*) and the hairy big-eared bat (*Micronycteris hirsuta*). Their calls attenuate quickly but allow for accurate pinpointing of obstacles and prey at close range.

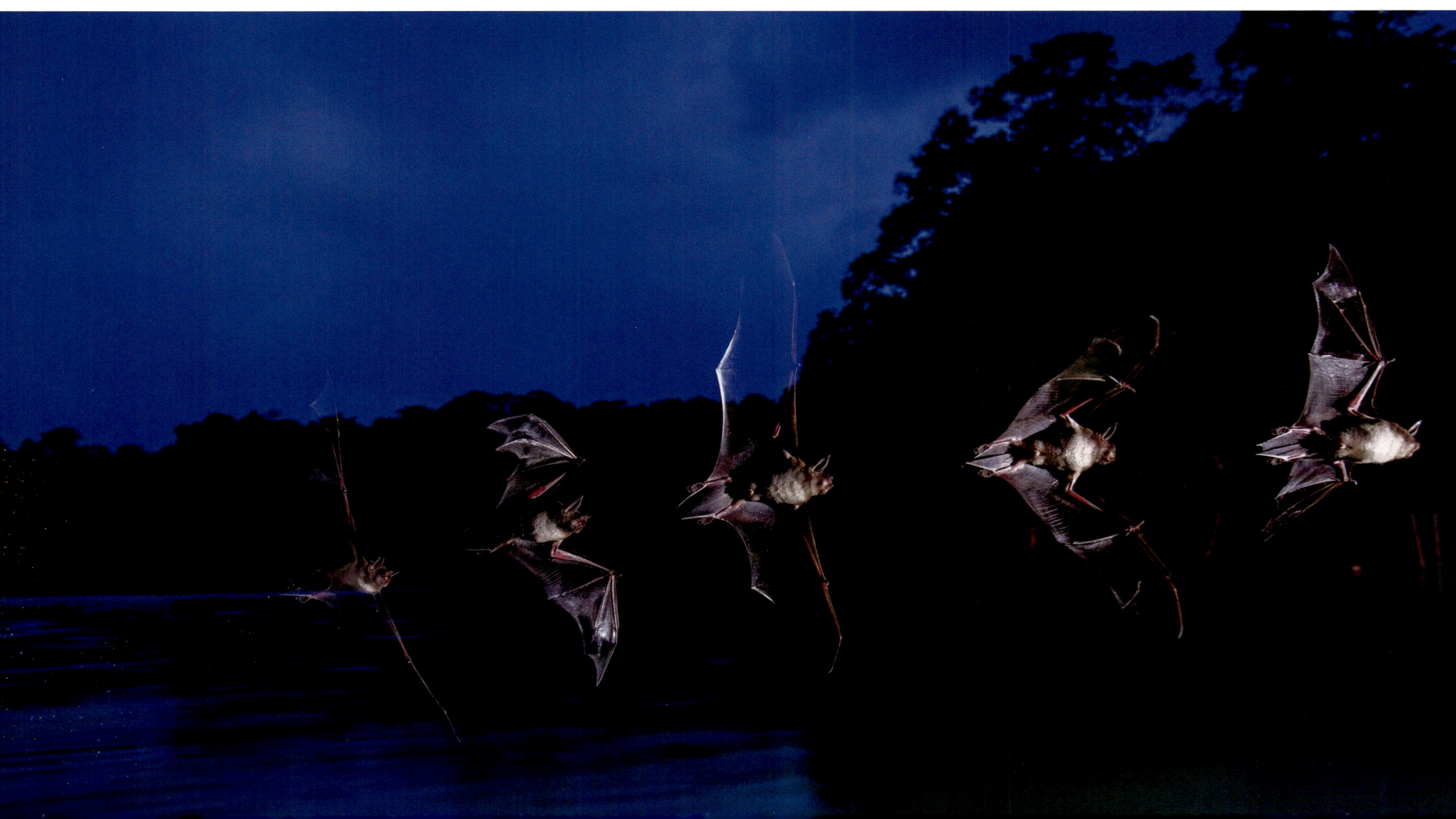

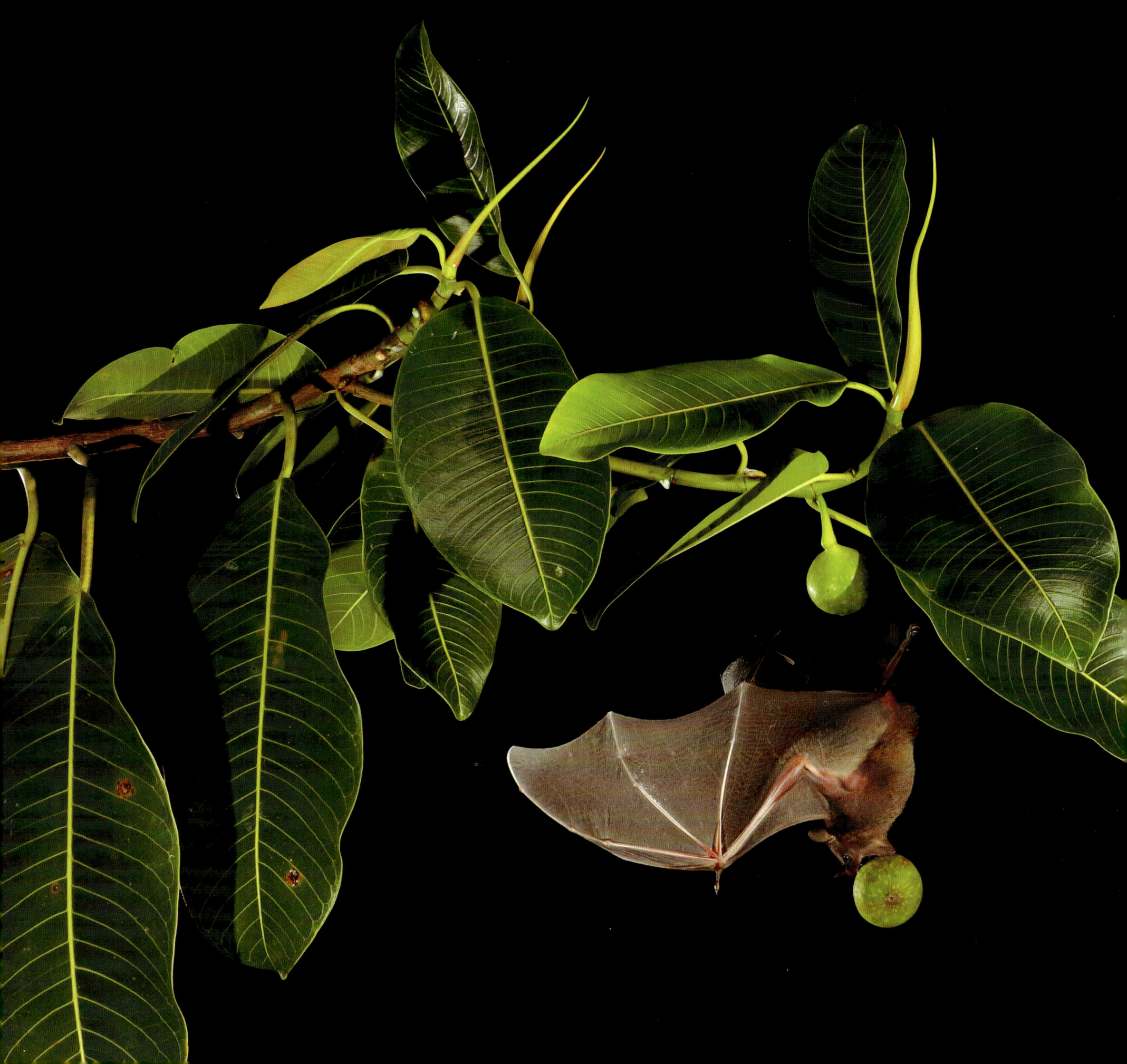

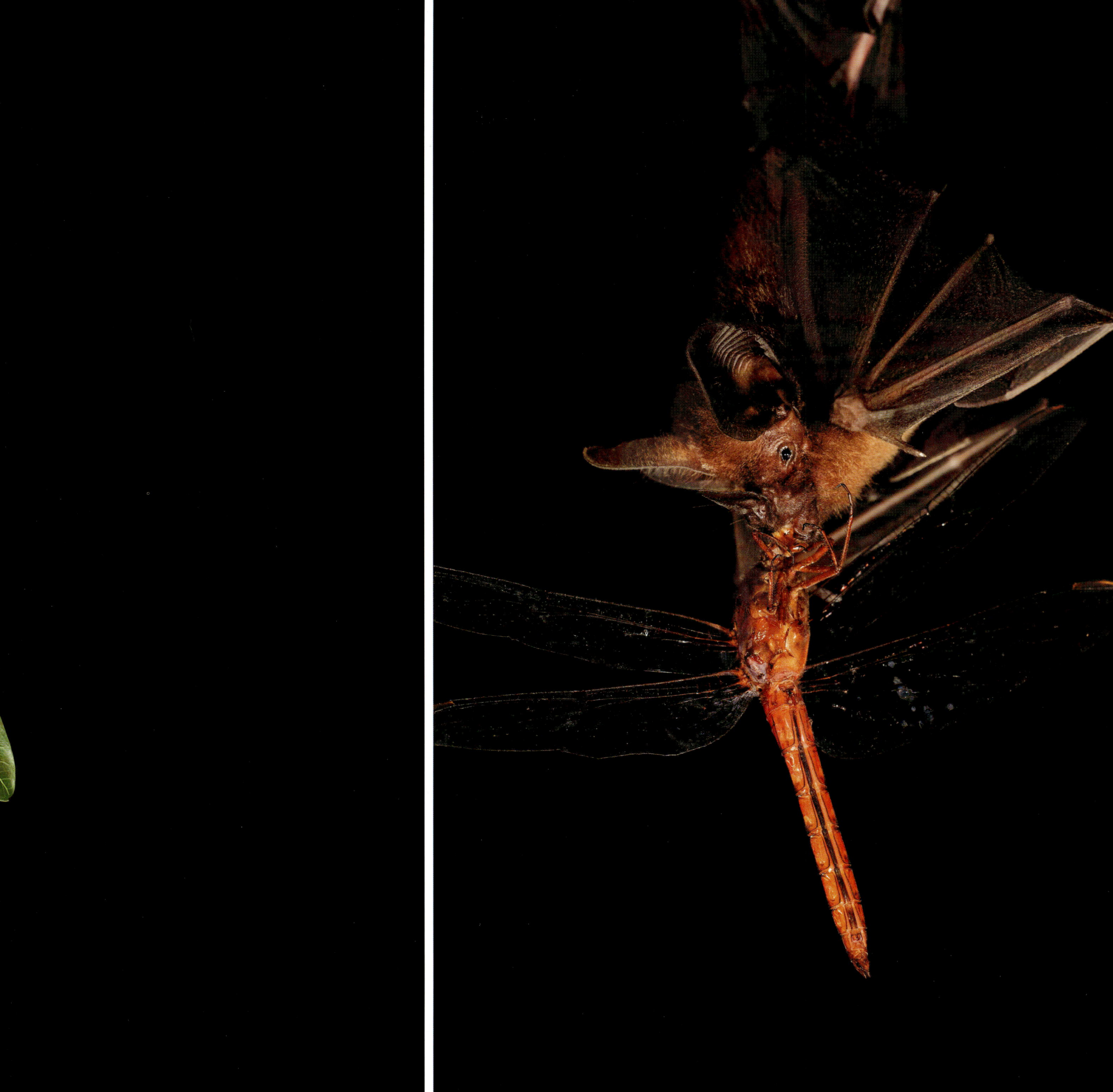

Bats do not use echolocation only for navigation and finding insect prey. Particularly in the tropics, bats forage on a wide variety of foods, including fruits and flowers. Nectar-feeding bats can use echolocation for detecting flowers, and frugivorous bats can use it to recognize fruit (see chapters 8 and 9). While temperate zones are dominated by bat species that feed on insect prey, which they detect primarily through echolocation, in tropical rainforests there are many additional foraging opportunities for which echolocation alone does not provide bats with sufficient information. For example, in the dense understory it is difficult to discriminate target echoes—those from hanging fruit or a stationary katydid—from the background echoes bouncing off surrounding vegetation. In such situations, bats rely on other sensory modalities and flexibly switch among them, integrating the information they receive as the situation requires.

PAGE 48 A Jamaican fruit-eating bat (*Artibeus jamaicensis*) performs acrobatic maneuvering to grab a fig. Frugivorous bats use a combination of olfactory and echolocation cues to find ripe fruits and play critical roles in regenerating forests and maintaining healthy ecosystems by dispersing seeds. **PAGE 49** The common big-eared bat (*Micronycteris microtis*) uses sophisticated echolocation to find silent, motionless prey in the dense forest understory at night. Here, it can be seen consuming a dragonfly, prey that is active in the day and sleeps on vegetation at night. *M. microtis* are very small bats, weighing roughly 0.21 ounces (6 grams), but consume an impressive variety of prey, including insects longer than their body length. In addition to dragonflies, they feed on caterpillars, katydids, spiders, beetles, cockroaches, stick insects, and even lizards they find sleeping on the vegetation, making them the smallest vertebrate-eating bat in the world. **BELOW** A fringe-lipped bat (*Trachops cirrhosus*) flies off after a successful túngara frog (*Engystomops pustulosus*) capture.

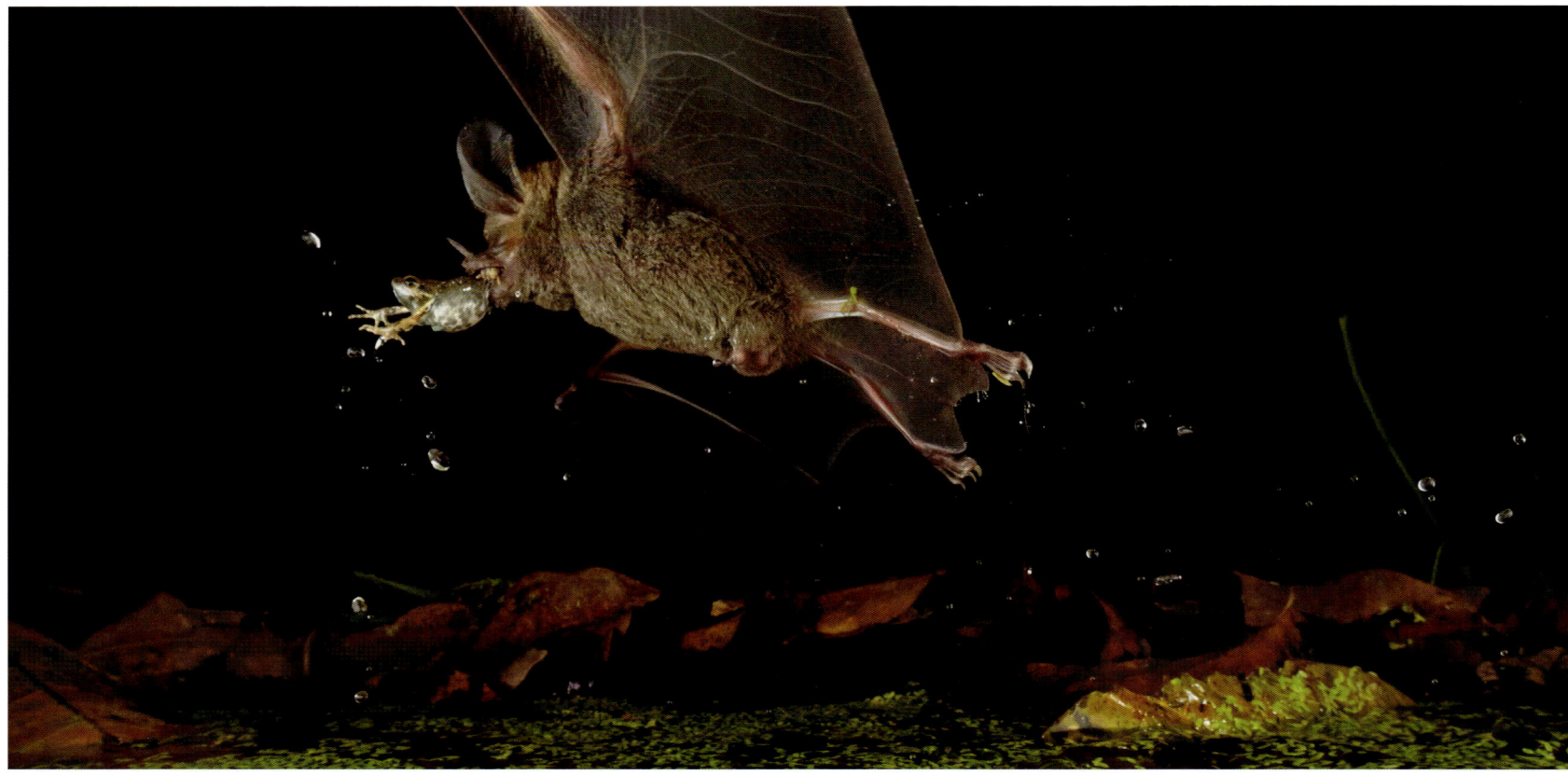

PASSIVE LISTENING

In contrast to echolocation, which is considered an active sense requiring a bat to invest energy into emitting sound to probe its environment, bat species that forage in dense forest conditions often use passive listening. Instead of sending out echolocation calls to detect prey, passive listeners attend to the sounds their prey produce. These could be locomotion noises, such as the wing-beat sounds of a fluttering cicada, the rustling of a small rodent in leaf litter, or the sound of a katydid landing on a leaf's surface. The big-eared woolly bat (*Chrotopterus auritus*), for example, hunts small vertebrates by listening for the noises they produce as they move on the forest floor.

Some bats, however, have taken passive listening a step further and eavesdrop on the communication signals their prey produce. Many nocturnal animals use sound to advertise for mates. Male frogs, for example, congregate in large choruses, calling loudly for females. Katydids produce calling songs to attract the opposite sex. Several bat species of large-eared passive listeners, including the white-throated round-eared bat (*Lophostoma silvicolum*) and the stripe-headed round-eared bat (*Tonatia saurophila*), eavesdrop on singing katydids and use these advertisement calls to localize their prey.

Bats do not just listen to the sounds produced by their prey. Aerial insectivores can also eavesdrop on the feeding buzzes of other bats, which act as reliable indicators of prey capture attempts. Especially for bats that must locate highly dispersed patches of prey, such as ephemeral swarms of insects, keying in on these distinctive echolocation calls greatly extends the detection range of an individual bat. Often the acoustic signal emitted by prey is the first beacon that alerts the predatory bat from a distance. As the bat gets closer, it can switch to its other sensory modalities, including olfaction and echolocation, to assess and localize prey.

The large ears of a big-eared woolly bat (*Chrotopterus auritus*) listening for the telltale sounds of prey. Photo © Marco Tschapka.

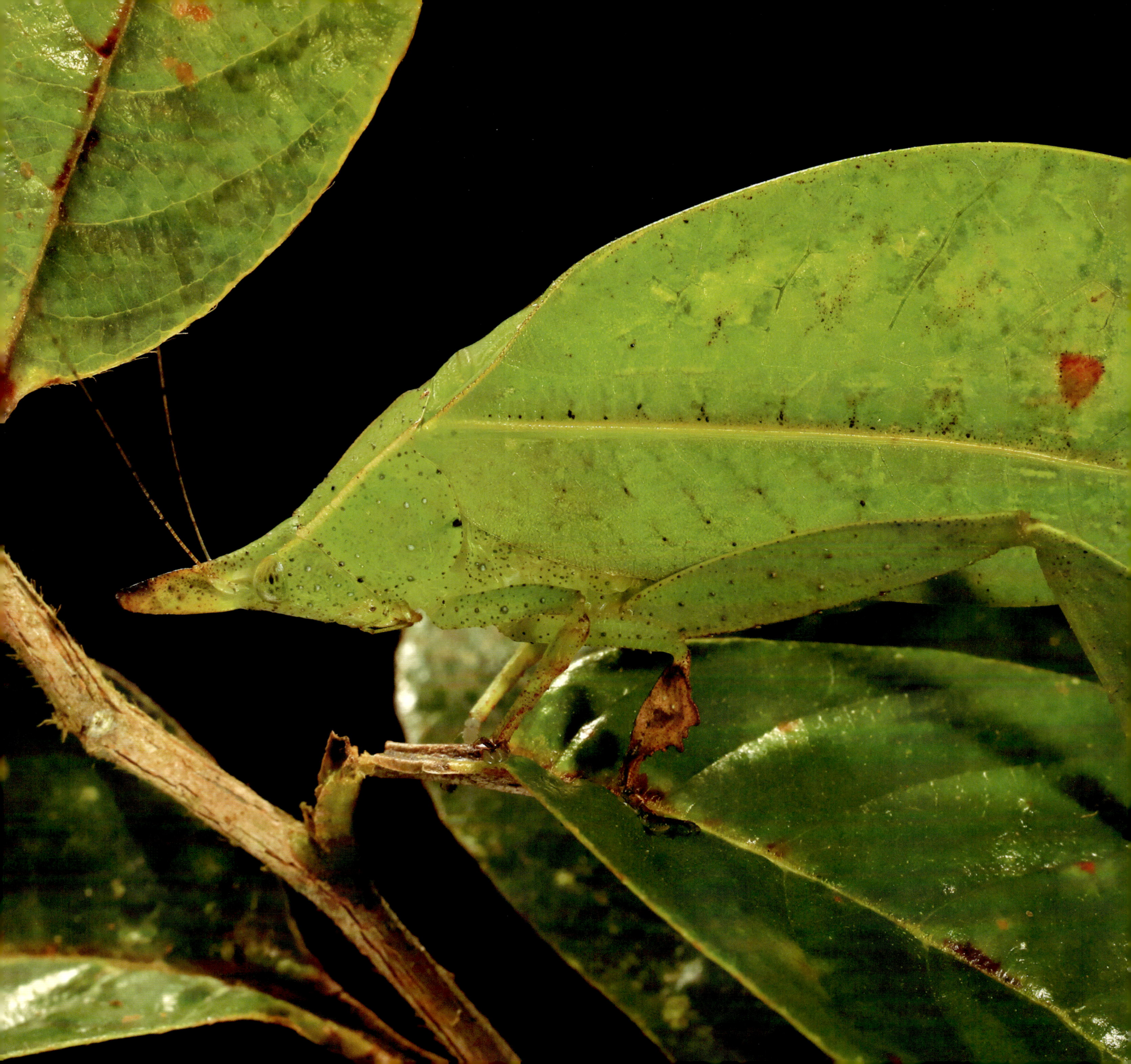

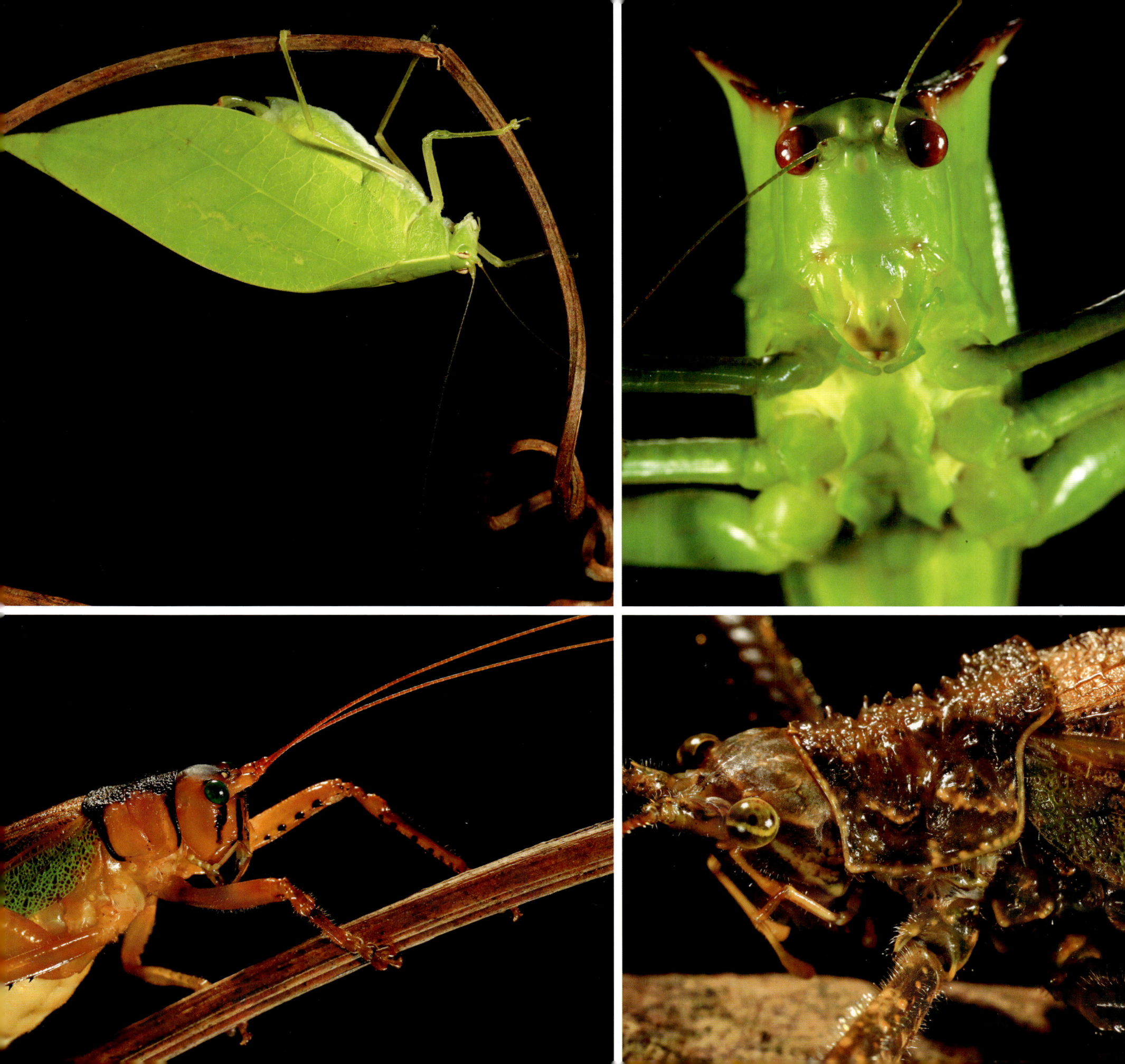

OLFACTION

As in all mammals, scent is extremely important to bat social behavior and plays a key role in courtship and reproduction. Bats often advertise who they are via scent, which is emitted from a variety of glands and other odor-producing structures, with odorous secretions smeared over various parts of the body, depending on the species. Reproductive male fringe-lipped bats (*Trachops cirrhosus*), for example, systematically apply a golden, odorous crust to their forearms. Several species of Neotropical ghost bats (*Diclidurus* spp.) exhibit a glandular area on the uropatagium—the membrane that stretches between the bat's hind limbs—which is prominent in males during the mating season. Male bulldog bats (*Noctilio albiventris* and *Noctilio leporinus*) have so-called inguinal organs, striking projections of skin that protrude when the testes are seasonally enlarged, spreading an intense garlicky smell. Most strikingly, perhaps, are male greater sac-winged bats (*Saccopteryx bilineata*), which ferment a strange concoction of urine, saliva, and gular glandular secretions in the pouches on their wings. With distinctive hovering flights, they waft the smell of this appealing cocktail at females in their harems, as well as at intruding males.

Bats' excellent sense of smell is not only used for olfactory communication within a species, but also for finding food. Bat-pollinated flowers use volatile chemicals to advertise their presence to flower-visiting bats, and bat-dispersed fruits are often highly odorous as well (see chapters 8 and 9). Bats can follow these odor plumes over large distances, and then use echolocation in the final stage of their approach to pinpoint the target.

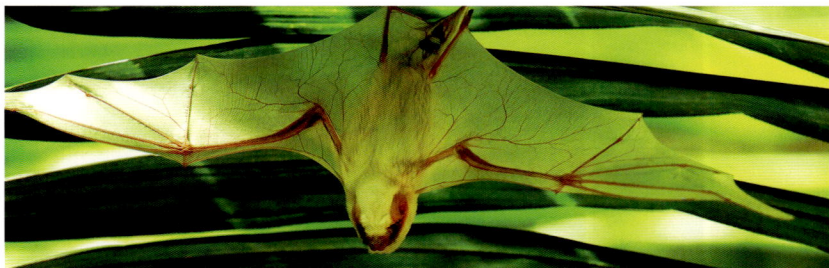

VISION

While sound—whether echolocation or passive listening—and smell are the most widely used senses in most bats, some bat species also rely heavily on other sensory modalities, including vision. With the exception of a few bat species that echolocate using tongue clicks or wing clapping, the entire family of Old World fruit bats, Pteropodidae, does not echolocate at all. All species from this group have large eyes and rely heavily on vision. In contrast, all New World bats do use echolocation, but several species also have very large eyes. For example, as the name suggests, big-eyed bats (*Chiroderma villosum*) on Barro Colorado Island have very large eyes and are thought to fly above the forest canopy, where they may encounter sufficient light to use vision to navigate among the treetops. The rarely captured wrinkle-faced bat (*Centurio senex*) is another species with huge eyes, but their role in this bat's ecology remains a mystery.

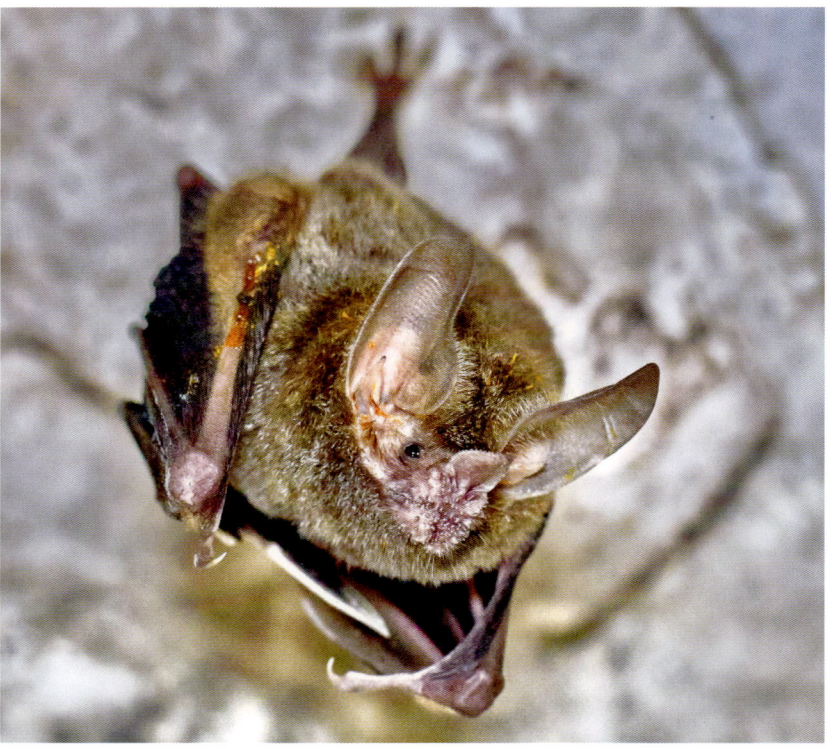

THERMORECEPTION, MAGNETORECEPTION, AND SOMATOSENSORY PROCESSING

In addition to the well-known sensory modalities of sight, smell, and sound, bats possess several lesser known but striking sensory abilities. The common vampire bat (*Desmodus rotundus*), for example, has specialized thermosensors near its nose that enable it to perceive heat. These organs allow vampire bats to find blood vessels close to the skin's surface of their hosts and feed efficiently. Some bat species even use magnetoreception, which allows them to use the Earth's magnetic field when navigating long distances or in unfamiliar terrain. Somatosensory hairs on bats' wing membranes provide sensory-motor feedback and have been shown to play an important role in the regulation of flight. As our tools for observing and studying bats become ever more sophisticated, more doors are opening in our understanding of a bat's sensory world, bringing us closer to answering the question Nagel posed half a century ago: What is it like to be a bat?

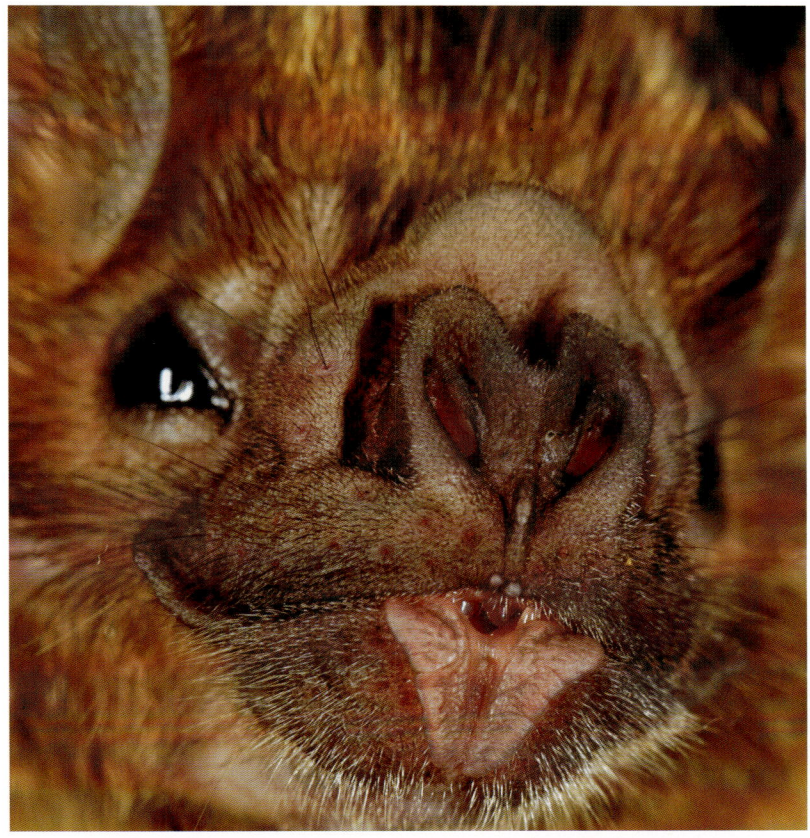

PAGES 52–53 BCI hosts over eighty species of katydids, many of which end up in the stomachs of hungry bats. Many show remarkable camouflage, with wings closely matching both the color and pattern of forest leaves such as *Aegimia maculifolia* (page 52) and *Viadana brunneri* (top left page 53). Others evade predators through armor, with formidable spikes and spines like *Steirodon careovirgulatum* (top right page 53) and *Pristonotus tuberosus* (bottom right page 53), while others are inconspicuous but have a strong bite, like *Cocconotus wheeleri* (bottom left page 53). **OPPOSITE LEFT** During the breeding season, male northern ghost bats (*Diclidurus albus*) exhibit a dark uropatagial sac between their hind limbs. **OPPOSITE RIGHT** Reproductive male fringe-lipped bats (*Trachops cirrhosus*) apply a pungent golden crust to their forearms; males with higher concentrations of testosterone exhibit larger forearm crusts. **TOP RIGHT** Common vampire bats (*Desmodus rotundus*) have heat-sensing organs near the nose to help them locate accessible blood vessels on their hosts. **BOTTOM RIGHT** Hairy big-eyed bats (*Chiroderma villosum*) have large eyes that may allow them to navigate above the forest canopy using vision. Opposite left photo © Manuel Sánchez Mendoza. Opposite right photo © Paul B. Jones. Top and bottom photos © Marco Tschapka.

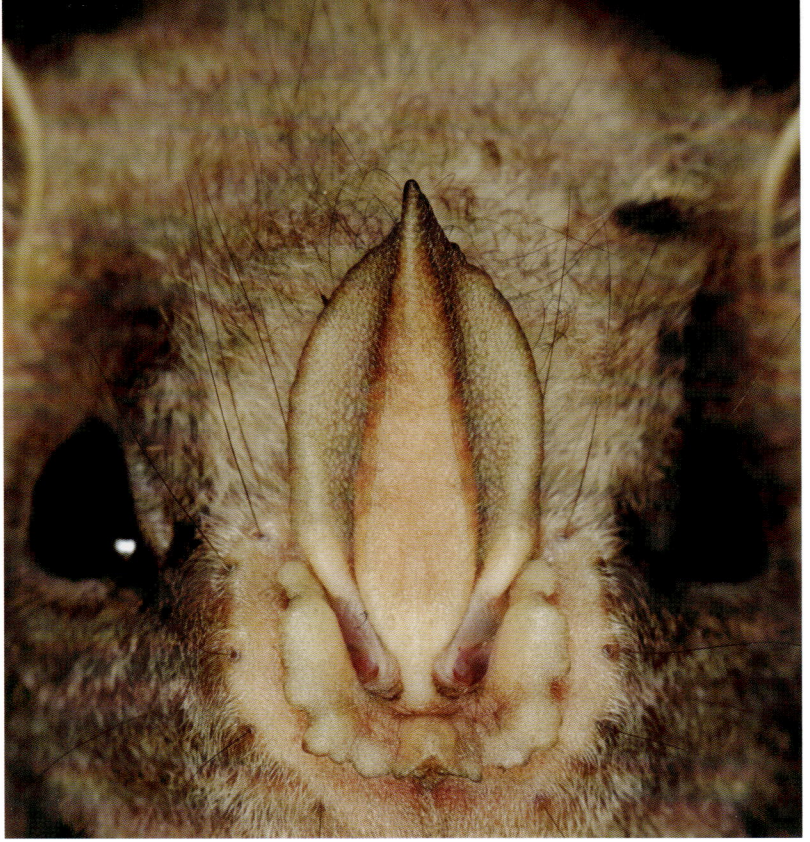

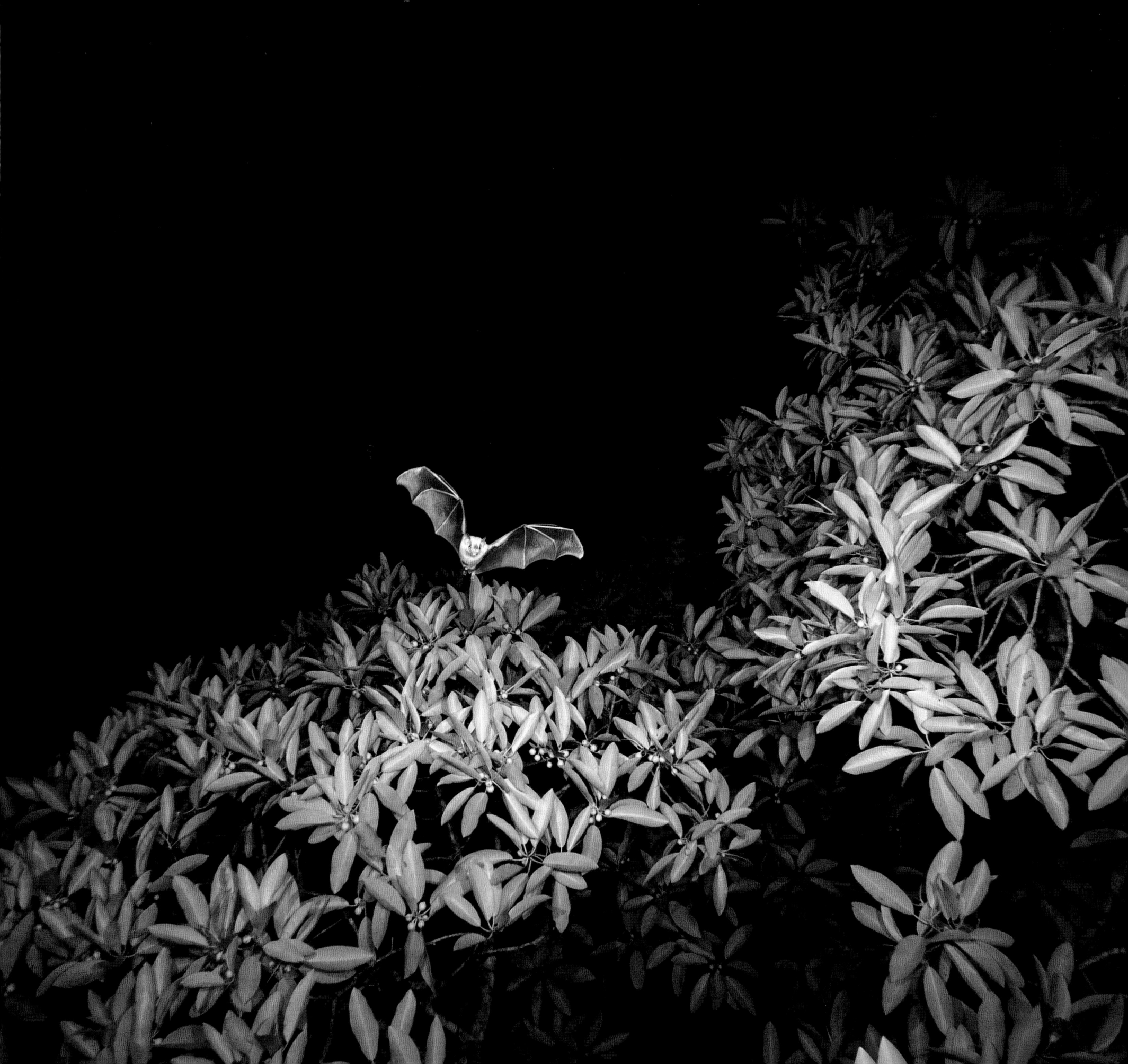

CHAPTER FOUR

CURIOSITY SERVES THE BAT

FLEXIBILITY, LEARNING, AND MODE SWITCHING

In addition to their extraordinary sensory abilities, bats have an array of cognitive adaptations that have contributed to their success and diversification. Bats are curious, quick learners. They rapidly acquire new information from both their own individual experiences and by observing the behavior of other bats. They have long memories, are highly flexible in their use of available cues, and navigate challenging sensory tasks by rapidly switching between relevant sensory modalities. Their long lifespans and complex social lifestyles may have contributed to selective pressures for evolving these advanced cognitive skills.

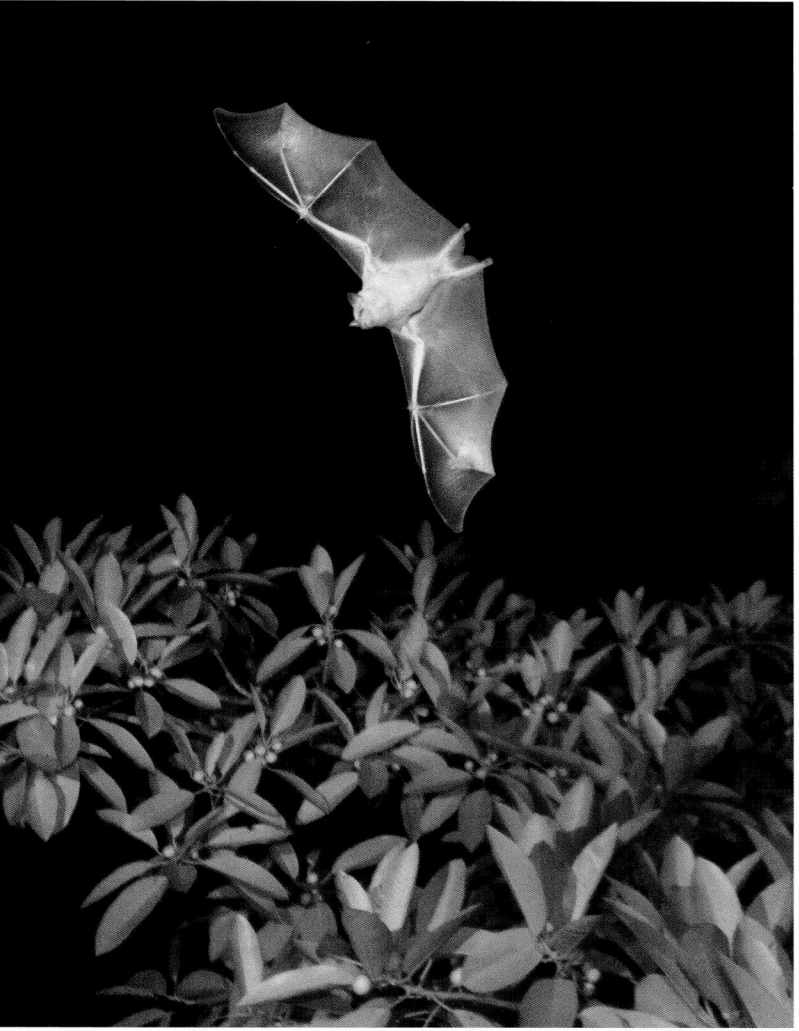

OPPOSITE AND RIGHT A Jamaican fruit-eating bat (*Artibeus jamaicensis*) flies in to inspect figs (*Ficus* sp.) On BCI, there are a number of fig-eating bat species of varying body size. Bats use a combination of scent and echolocation to find and identify fig species, and prefer figs that match their body size. Figs on BCI range from nearly 1 inch (3 centimeters) in diameter down to below 0.39 inch (1 centimeter) in diameter. Larger frugivorous bats predominantly feed on larger figs, while the smaller frugivorous bats can only handle smaller figs.

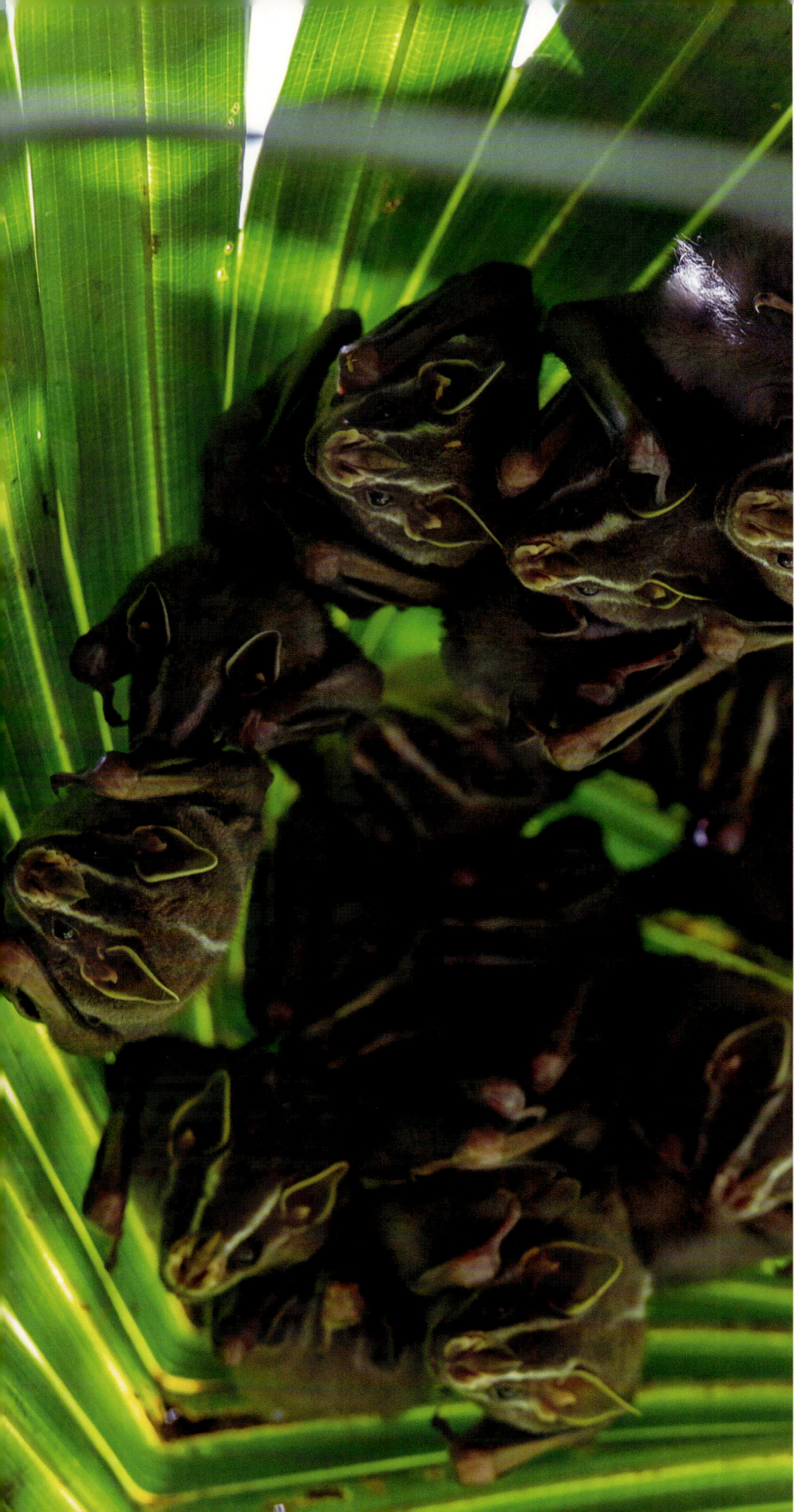

LONG LIVES

For their size, bats lead extraordinarily long lives. Across all mammals, longevity generally scales with body size: the larger the organism, the longer its life. Relative to their body size, only nineteen species of mammals are known to be longer lived than humans, and eighteen of these species are bats. Other mammals with a similar body size to bats, such as shrews and mice, generally live just over one year. The longevity record for a bat is forty-one years, held by a 0.2-ounce (6-gram) cave-dwelling Brandt's bat (*Myotis brandtii*) found in Siberia. It is thought that the anti-inflammatory adaptations that help bats cope with the metabolic by-products generated from powered flight may have also conferred on bats both an excellent immune system and a long lifespan. Bats' low reproductive rate—having just one or in some cases two pups per year—is another selective force that may have contributed to bats' long lives. While short-lived animals ensure that genes are passed to the next generation by having large litters with many offspring, flight constrains litter size in bats. Mothers can fly with one or at most two pups at a time; their reproductive phase is thus stretched out over years. Given their extraordinary longevity, it would make sense for bats to have long-term cognitive skills and memory to help them retain valuable foraging information, remember roost sites, and navigate multiyear social partnerships. Research over the last decades has shown this to be the case.

It is thought that bats' extreme longevity–highly unusual for mammals this small–allows for investment in advanced cognitive skills. Peters' tent-making bats (*Uroderma bilobatum*) are quick social learners. Their roosts serve as information centers; bats decide which fruits to feed on by attending to the smells brought back to the roost by recently fed bats.

FLEXIBILITY AND QUICK LEARNING

The fringe-lipped bat (*Trachops cirrhosus*) is an excellent example of flexibility and quick learning in bats. This species feeds on a wide variety of animals, including frogs. It homes in on the advertisement calls male frogs use to attract their mates. Not only does *T. cirrhosus* use frog mating calls for prey detection and localization, but it also uses these signals to determine whether a frog is palatable or poisonous, and whether it is small enough to eat or large enough that it could turn the tables and eat the bat—information gleaned solely from the frog's call. Behavioral experiments with *T. cirrhosus* show that these bats quickly learn to associate novel calls with palatable prey, and can rapidly redefine existing associations between the calls they hear and the expected prey quality. These bats are so flexible that early experiments showed that they can even learn to associate completely novel acoustic cues, such as the music of Bob Marley or cell phone ringtones, with the presence of prey.

Part of the reason that these bats are such quick, flexible learners is that they can afford to make mistakes. They have multiple mechanisms for assessing prey, such that a mistake at one stage of the prey assessment process can be corrected at another. *T. cirrhosus* uses frog calls at a distance as beacons of palatable prey, but in close proximity, its echolocation can tell it if the prey is an appropriate size to eat, or so large that it might turn the tables and consume the bat. At the last stage of prey capture, *T. cirrhosus* is thought to use the distinctive fringes on its chin and lips to further localize and assess its prey. When it comes in contact with the prey, it will reject prey that is coated in toxic secretions. *T. cirrhosus* also flexibly shifts its reliance on these different sensory cues depending on the environment. In noisy conditions, for example, when trying to pinpoint a single calling frog in the midst of the cacophony of a tropical night frog chorus, *T. cirrhosus* will increase its reliance on cues from multiple sensory modalities—also using echolocation to recognize the frog's dynamically expanding vocal sac or ripples that emanate on the water surface—to more accurately locate its frog prey.

Fringe-lipped bats (*Trachops cirrhosus*) hunt using the sounds generated by their prey. One is seen here in a flight cage experiment approaching a speaker playing the mating call of a tasty frog.

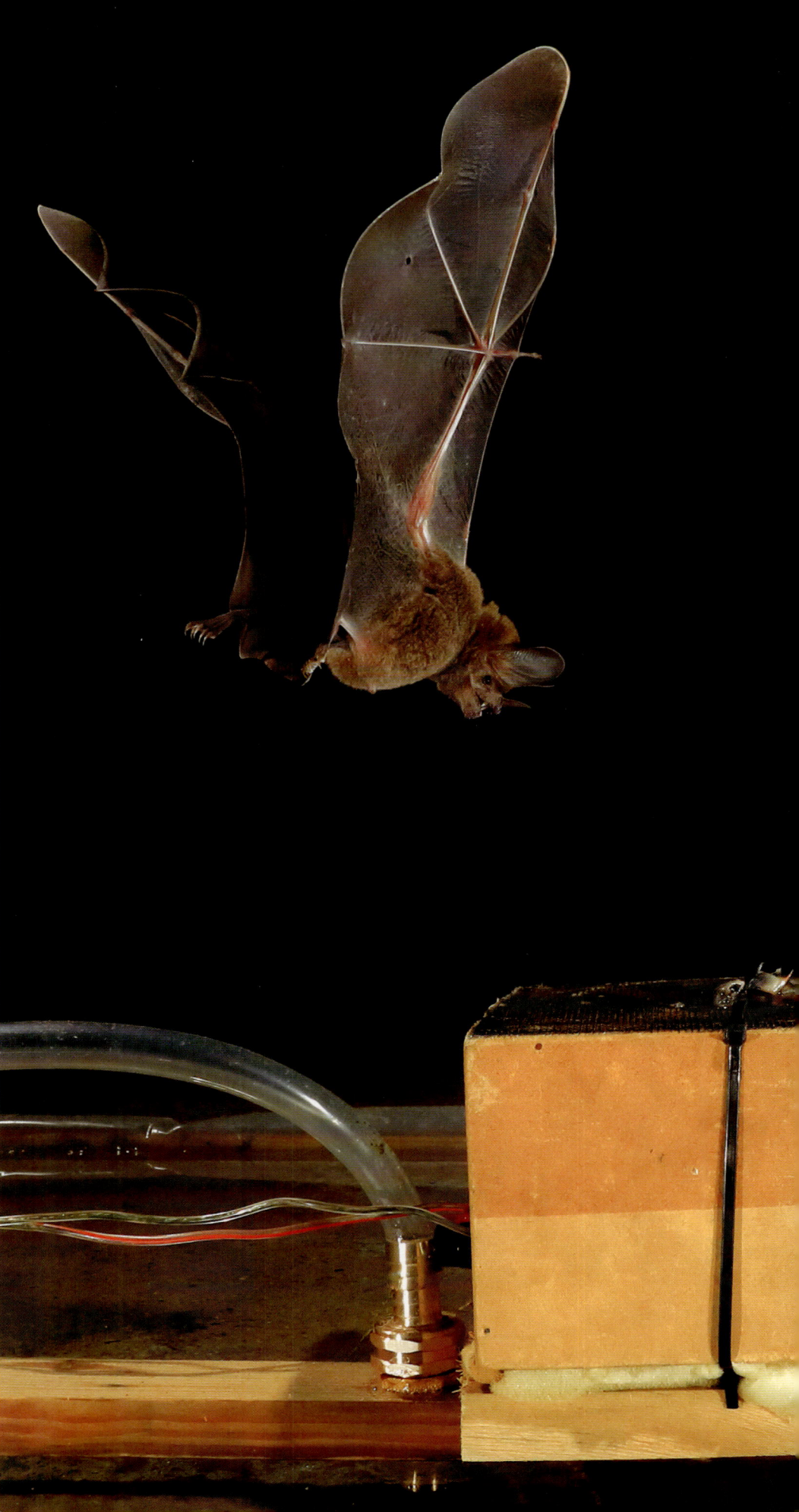

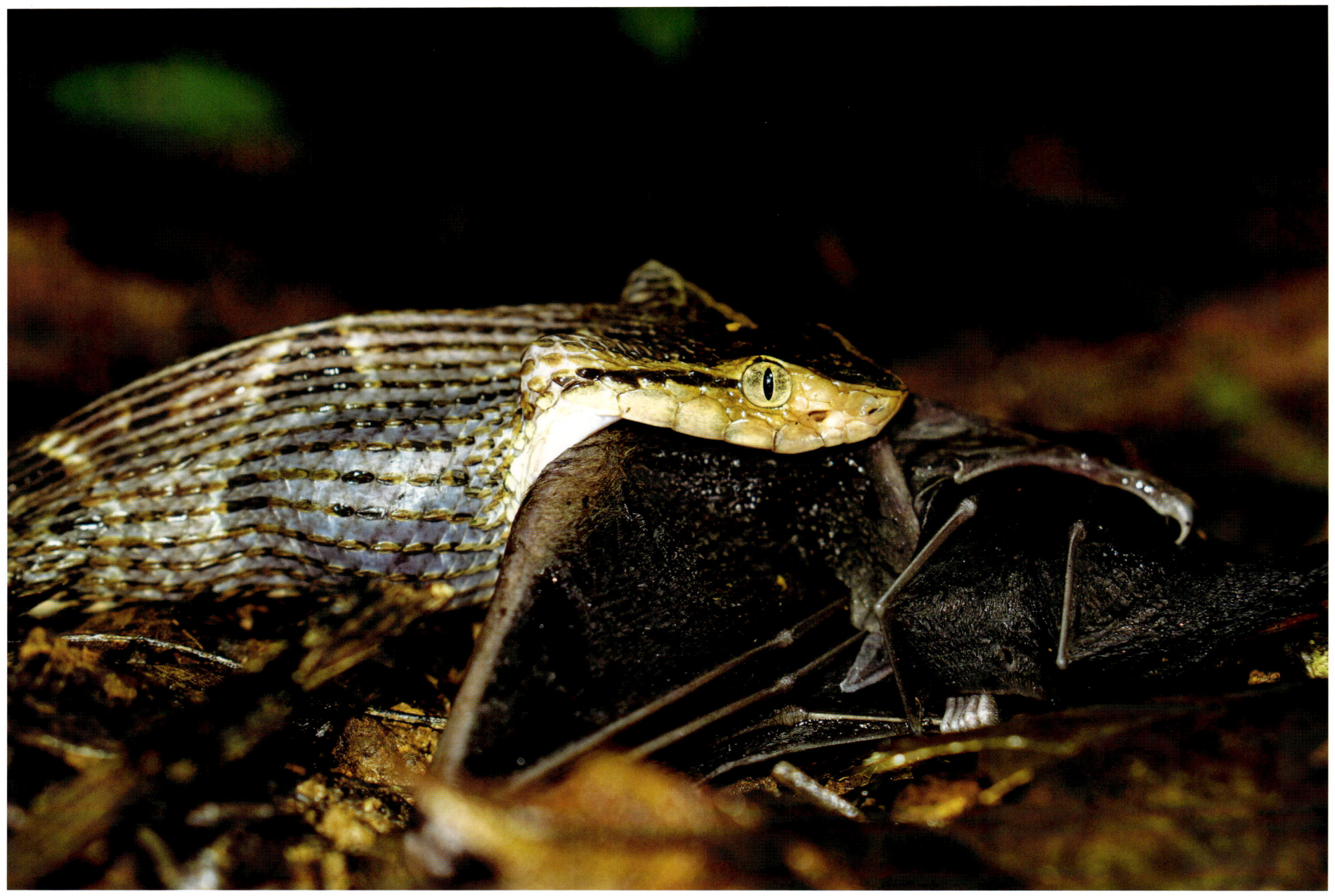

Fringe-lipped bats (*Trachops cirrhosus*) are attracted to ponds of chorusing túngara frogs. As they approach the water surface to take a meal, the tables are sometimes turned and the hunter can become the hunted. Here, a fer-de-lance (*Bothrops asper*) eats a bat that has approached a group of chorusing túngara frogs (*Engystomops pustulosus*). Photo © Hubert A. Szczygieł.

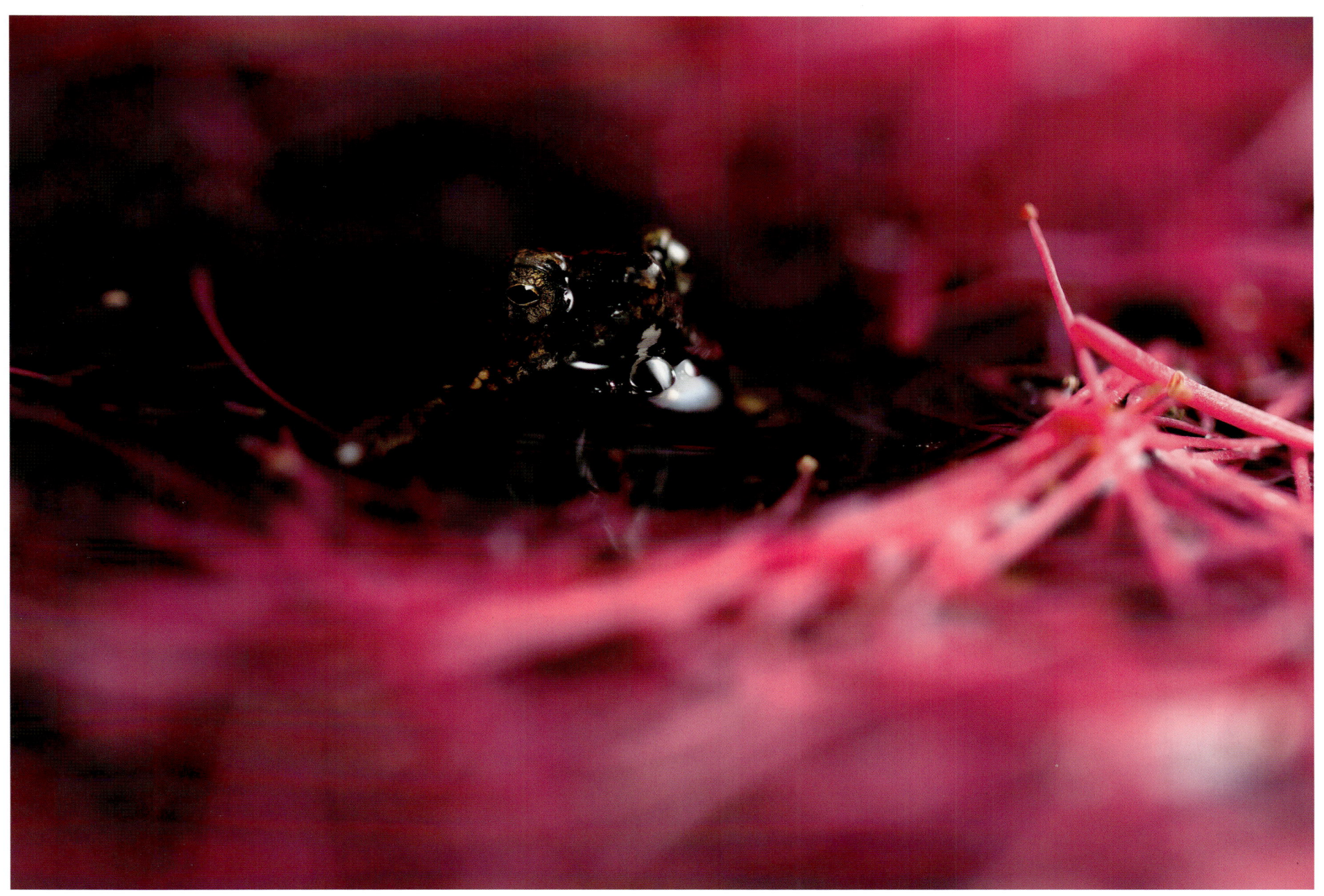

A túngara frog (*Engystomops pustulosus*) calls from a puddle full of flower petals from a marañón curasao, or water apple tree (*Syzygium malaccense*). Túngara frogs are one of the favorite foods of the fringe-lipped bat (*Trachops cirrhosu*s). Photo © Grant Maslowski.

SOCIAL LEARNING

Not only do bats learn quickly from cues they detect directly from the environment, but they are also capable of rapid social learning, both within and across species. Social transmission of learned information can take place through direct observation of another individual. The nectar-feeding Pallas's long-tongued bat (*Glossophaga soricina*), for example, rapidly learns new flower locations when able to observe the behavior of experienced individuals. Likewise, *T. cirrhosus* can learn to associate a sound it has never heard before with palatable prey by observing the hunting behavior of either another fringe-lipped bat or even the white-throated round-eared bat (*Lophostoma silvicolum*), a bat species of similar size that forages in the same areas and overlaps with *T. cirrhosus* in prey preferences.

Direct observation of another bat's foraging behavior is not necessary for social transfer of information to take place, however. Frugivorous bats are faced with the challenge of locating trees with ripe fruit, a demanding task in the expanse of a tropical rainforest. A bat that returns to the roost smelling of a particular type of fruit may influence the foraging decisions of others in that roost, essentially turning the roost into a place of information exchange. Because large fruiting trees are hard to find—but once found offer abundant resources that cannot be monopolized by an individual bat—there should be no cost in attracting roostmates to a discovered food source. Both Peters' tent-making bat (*Uroderma bilobatum*) and Seba's short-tailed bat (*Carollia perspicillata*) can socially acquire preferences for new foods through encountering the smell of fruits on roostmates that have recently eaten. Greater spear-nosed bats (*Phyllostomus hastatus*) may even recruit roostmates to flowering balsa trees with loud screech calls produced at the roost.

RIGHT A small tent-making bat (*Dermanura* sp.) peeks cautiously out of its roost in a leaf tent. Frugivorous bats such as these socially learn odor cues associated with foods consumed by returning roostmates. OPPOSITE A group of roosting greater spear-nosed bats (*Phyllostomus hastatus*). These bats use socially-learned, group-specific calls to stay in contact with roostmates when foraging for abundant resources such as flowering balsa trees (*Ochroma pyramidale*).

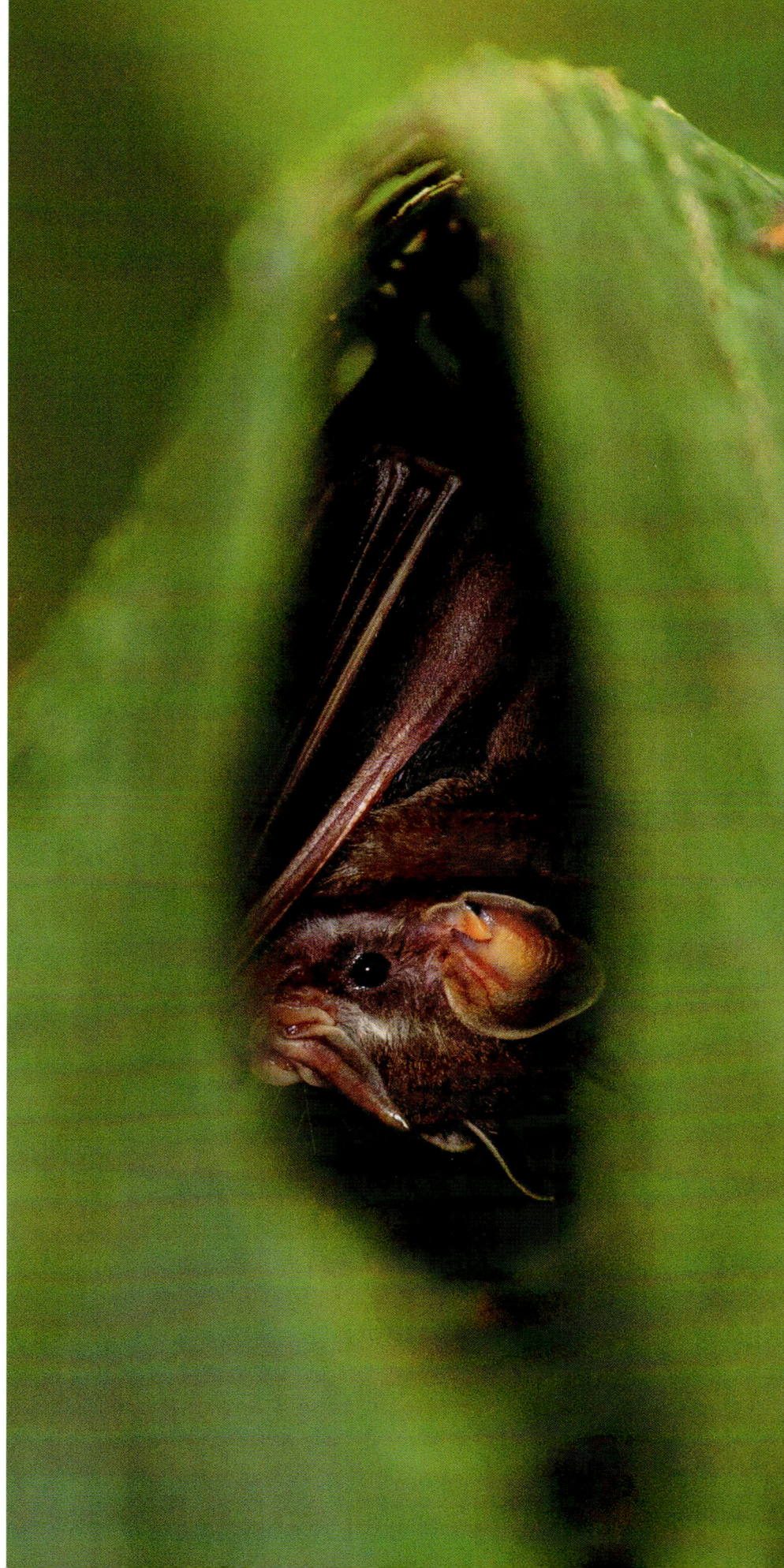

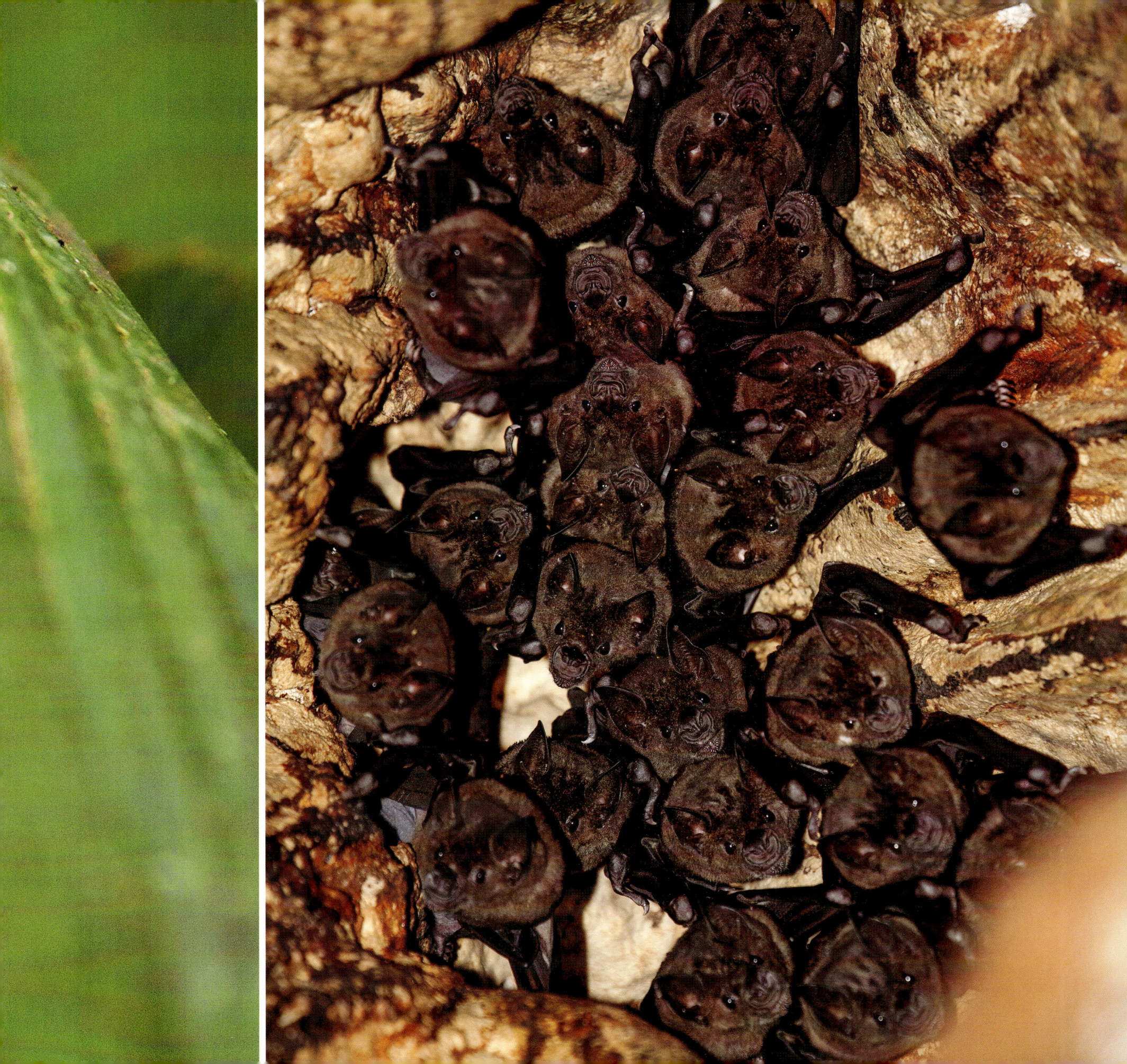

TEACHING

While the vast majority of social learning documented to date across the animal kingdom is passive—with one individual opportunistically reacting to the behavior of another—cases of active teaching have been documented, including in bats. The criteria for teaching include the following: An experienced individual (the teacher) must change its behavior in the presence of a naïve observer (the student); this change in behavior must not benefit, or could even be costly for the teacher; and the student must gain novel information from its interaction with the teacher. These criteria have been met in observations of the hunting behavior of the scorpion-eating pallid bat (*Antrozous pallidus*) living in arid regions of North America and seem likely to be present in many other bat species as well. While such interactions among bats are highly difficult to observe, especially in the wild, the most fruitful avenue for this research is likely among mother-pup pairs. In some bat species, mothers spend extended periods of time associating with their pups, even after weaning, which is likely when most social transfer of information takes place. The mothers of common big-eared bats (*Micronycteris microtis*) provision their pups with solid food for months after they stop nursing. They catch large katydids, stick insects, caterpillars, and even anole lizards, and bring them back to the roost to share with their pups. It has been hypothesized that provisioning within the roost not only provides the pups with valuable nutritional resources, but also teaches them the echo-acoustic images of profitable food to target once they begin to forage on their own. It may also teach them appropriate handling techniques, especially for noxious prey such as hairy caterpillars. Unlike closely related bats that use their thumbs, wings, and forearms to hold and manipulate their prey, *M. microtis* does not contact its prey with any body part except for the mouth, perhaps shielding it from noxious prey chemicals or spiny defenses.

Most bat species are highly social and roost in close body contact with one another other, such as the small 0.21-ounce (6-gram) common big-eared bat (*Micronycteris microtis*) at right, or the much larger 1.05-ounce (30-gram) fringe-lipped bat (*Trachops cirrhosus*), opposite. Mothers in some species such as these associate with their pups even after they stop nursing, allowing opportunity for social learning from mother to pup.

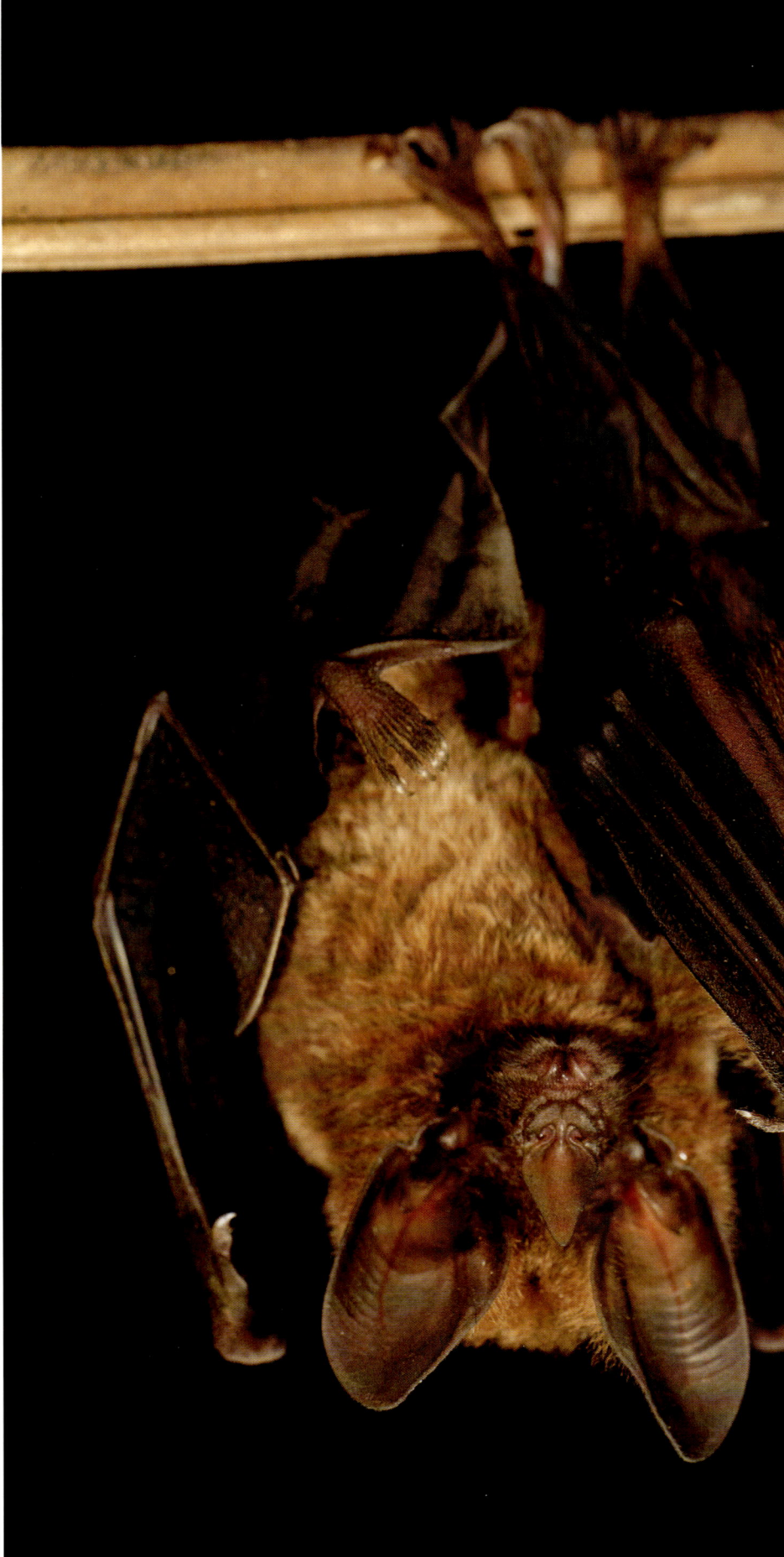

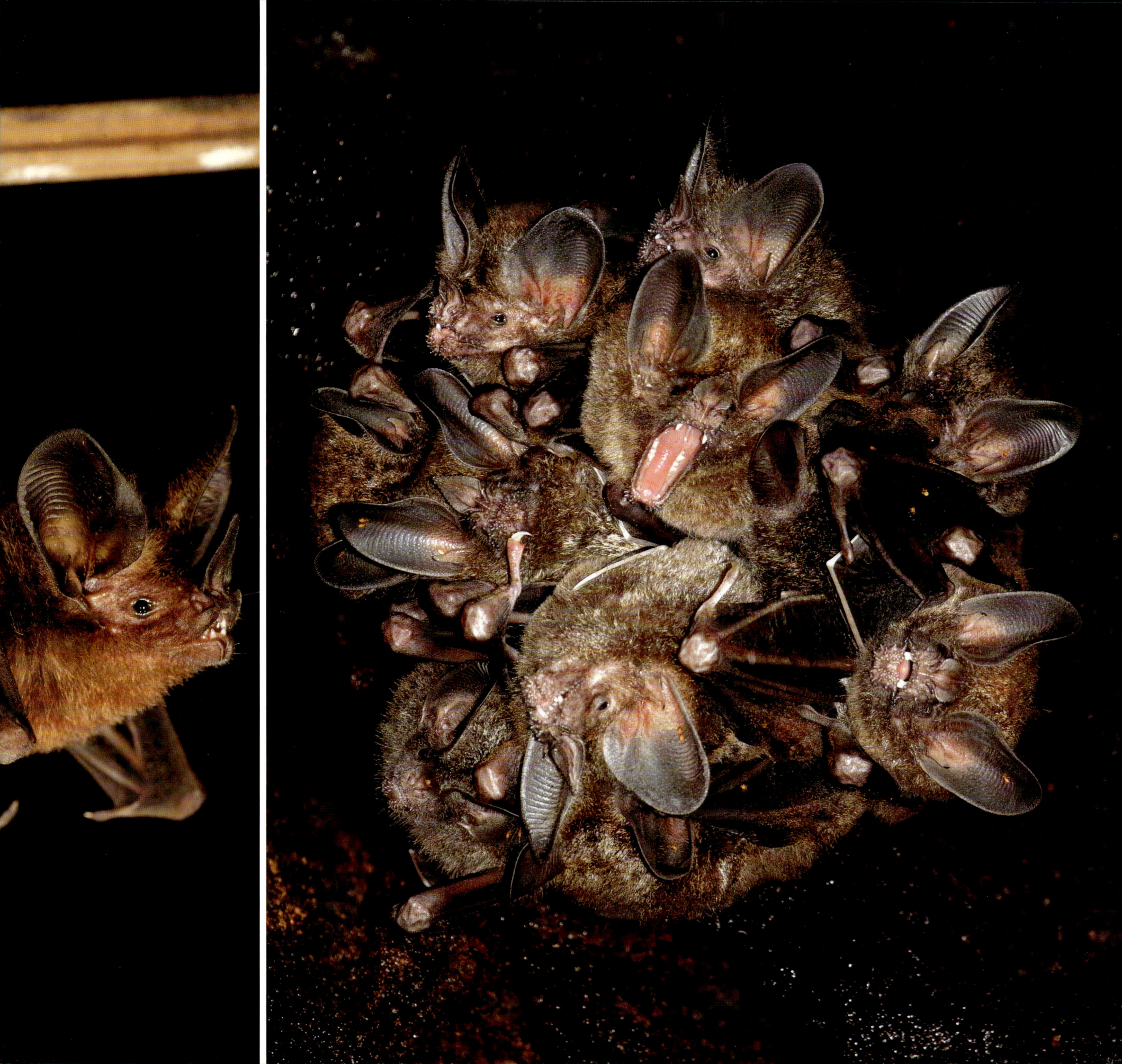

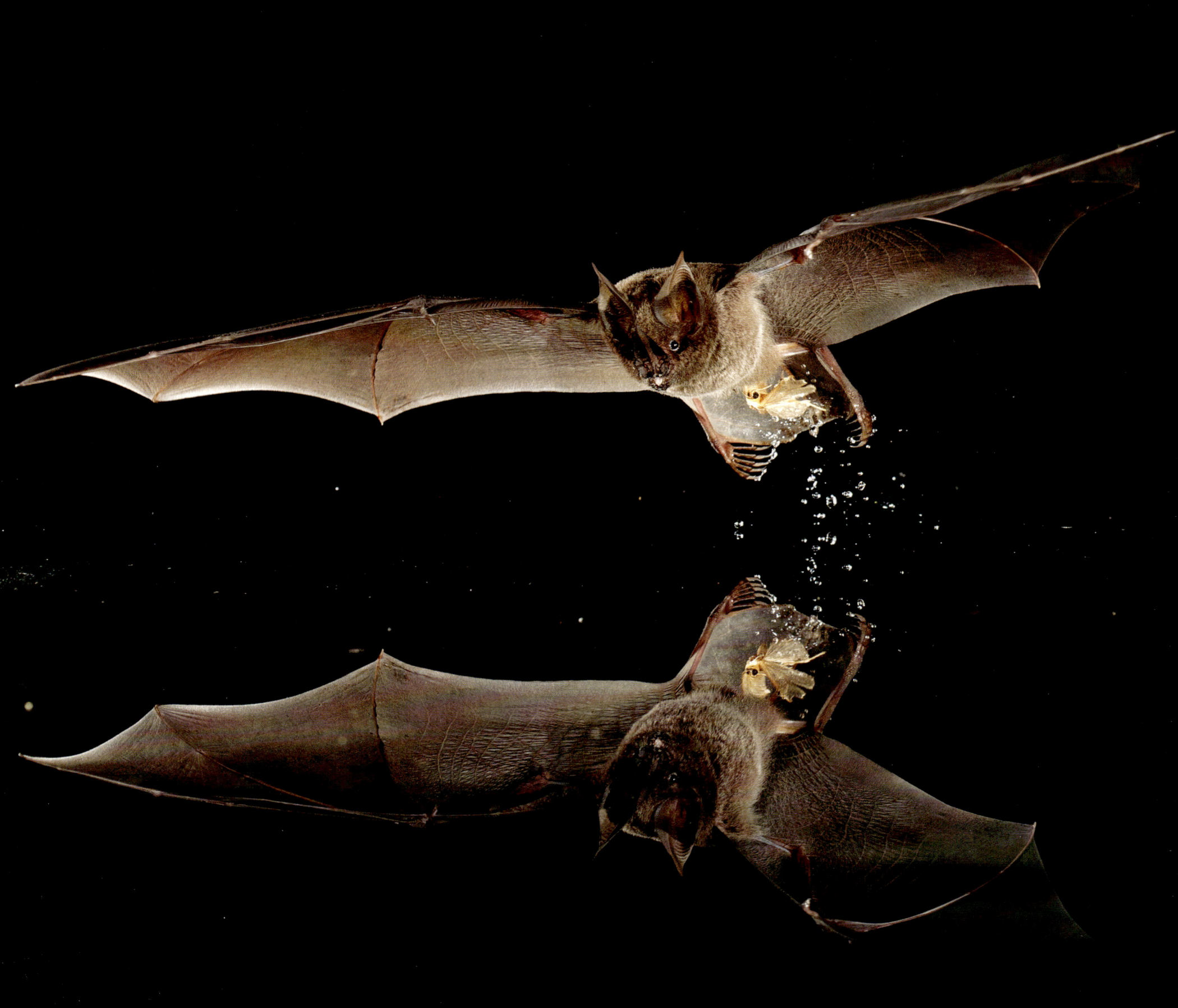

CHAPTER FIVE

TELL ME WHERE YOU FLY
AND I'LL TELL YOU WHO YOU ARE

THE GUILD CONCEPT

The term *ecological niche* describes the particular position and role a species occupies in its environment. An animal's unique suite of sensory and cognitive adaptations, social behaviors, roosting preferences, foraging abilities, predator evasion tactics, physiological tolerances, and many other traits combine to allow it to inhabit a distinct niche in its particular habitat. How bats evolved into different niches such that multiple species can coexist in one place—such as the seventy-six bat species on BCI—has been a driving question of evolutionary and ecological research. The study of animal community structure reveals that species fall into guilds, or functional groups that share similar traits. Foraging guilds, for example, are groups of species that find food in similar ways in similar parts of the landscape. Research carried out on BCI is pivotal not only in addressing this question, but also in identifying key characteristics that allow us to understand how bat communities are structured, what ecological niche a bat inhabits, and what foraging guild it falls into. Two of these characteristics are wing shape and echolocation.

WING SHAPE

A crucial discovery was made when research described how wing shape reflects what an individual is capable of performing while in flight. This fundamental aspect of a bat's biology, its flight style, determines many aspects of its ecological niche. Bats with long, narrow wings excel at fast flight in open spaces, but struggle to navigate dense forest understory. In contrast, bats with short, broad wings are highly maneuverable in tight spaces, but due to their slow flight would be vulnerable to fast-flying predators in open spaces. Similar predictive patterns are found in birds. An owl, for example, has broad, round, and, in comparison to its body size, large wings. From this wing shape we can tell that owls will not fly long distances regularly, but that they can maneuver extremely well in the understory and close to the ground, allowing them to pursue their prey of mice and other small mammals. Bats with a similar wing shape to owls share a similar foraging mode: They fly slowly in the understory and pick up their food, be it animals or fruit, from the ground or the vegetation. Conversely, fast-flying bats hunting aerial insects in open areas such as above the forest canopy have long and narrow wings and find their parallels in the bird world in swifts and swallows. In fact, wing shape is relevant to nearly every aspect of a bat's life, including its diet, foraging mode, and even sometimes its social system.

The long and broad wings of long-legged bats (*Macrophyllum macrophyllum*) allow them to expertly glean insects from the water's surface. **PAGES 68–69** A common vampire bat (*Desmodus rotundus*) leaving a hollow tree at the beginning of the night. Its broad wings allow it to maneuver through the forest understory and take off from the ground as it approaches its host for a blood meal. Photo © Marco Tschapka.

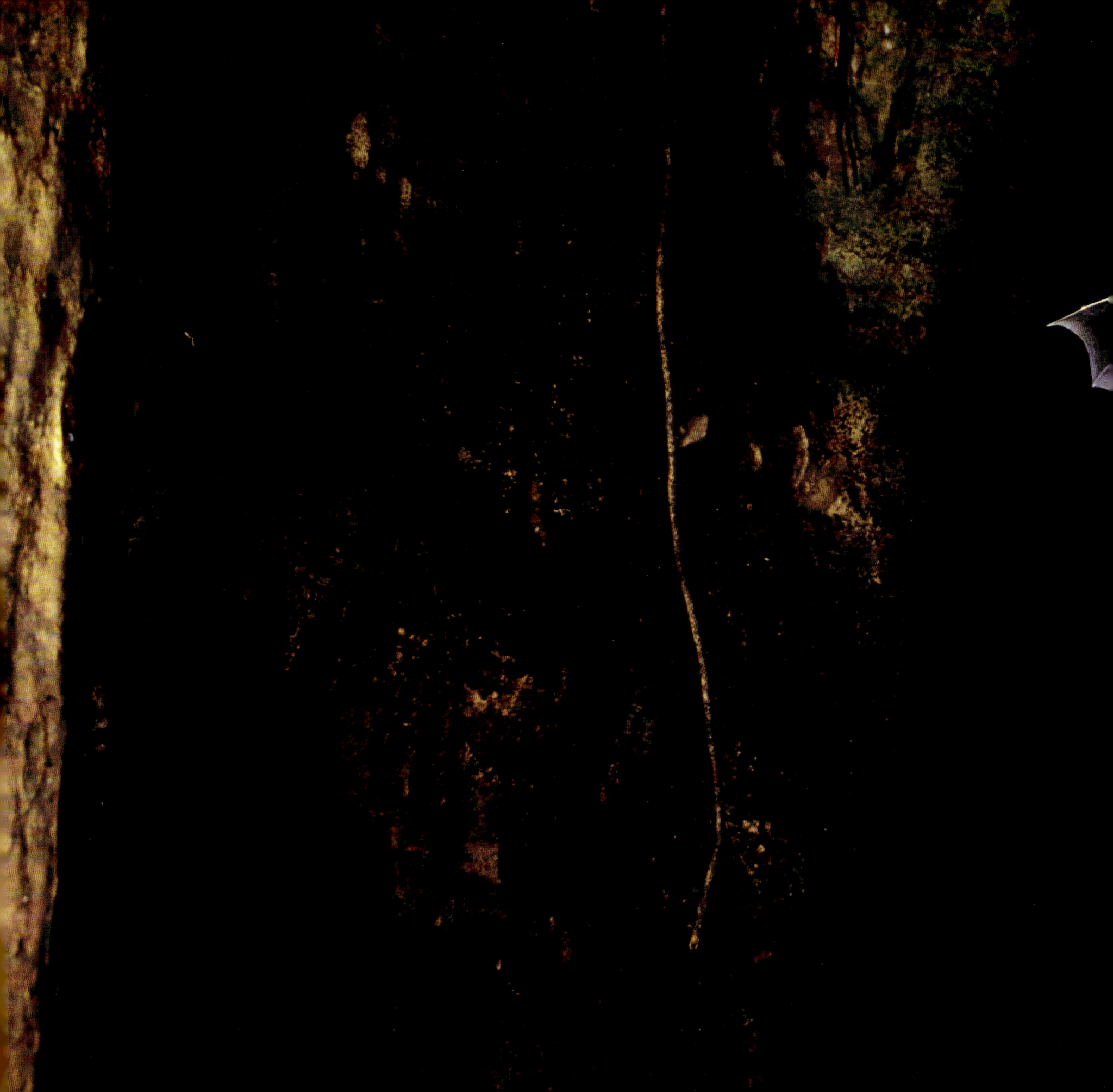

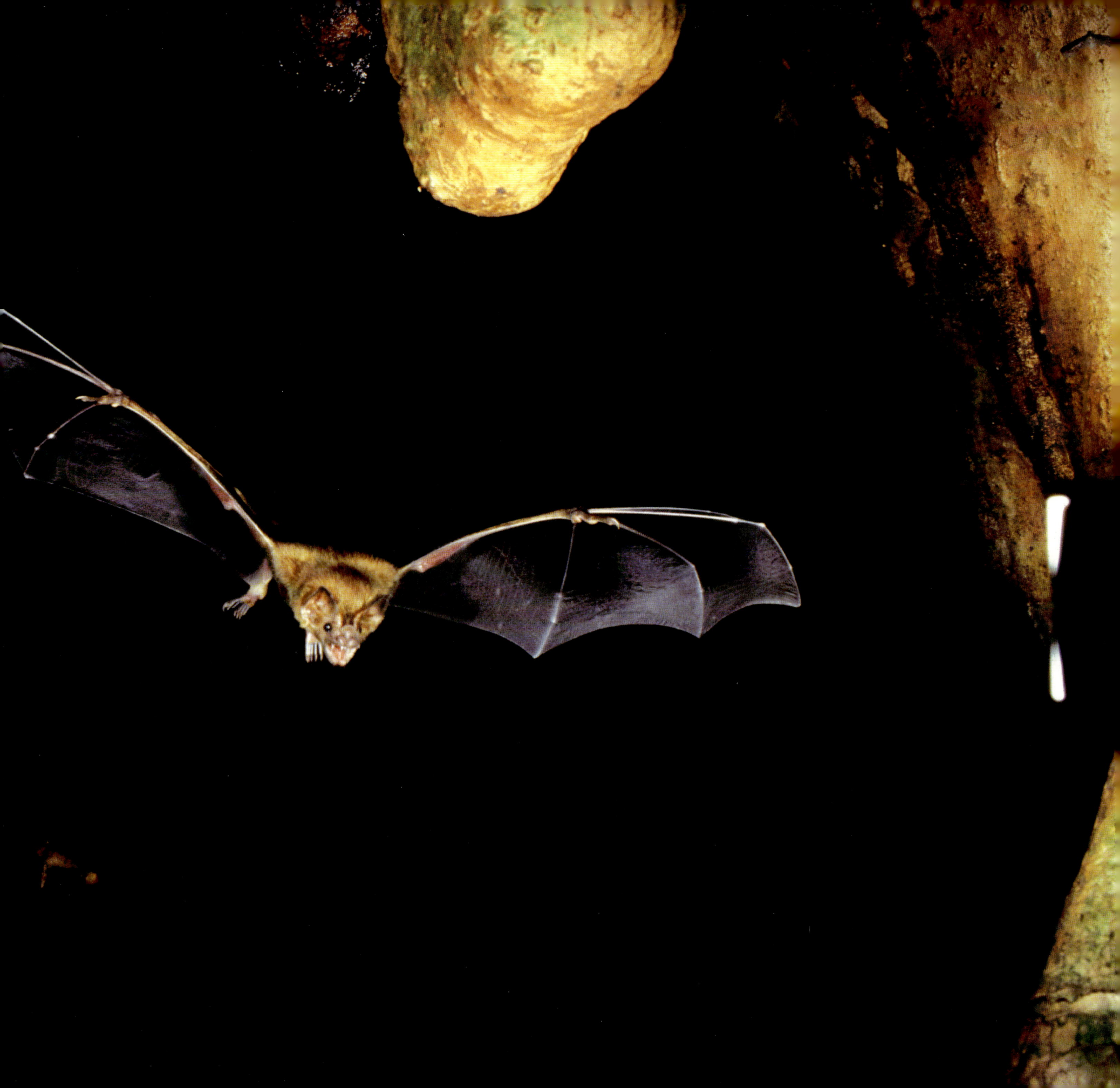

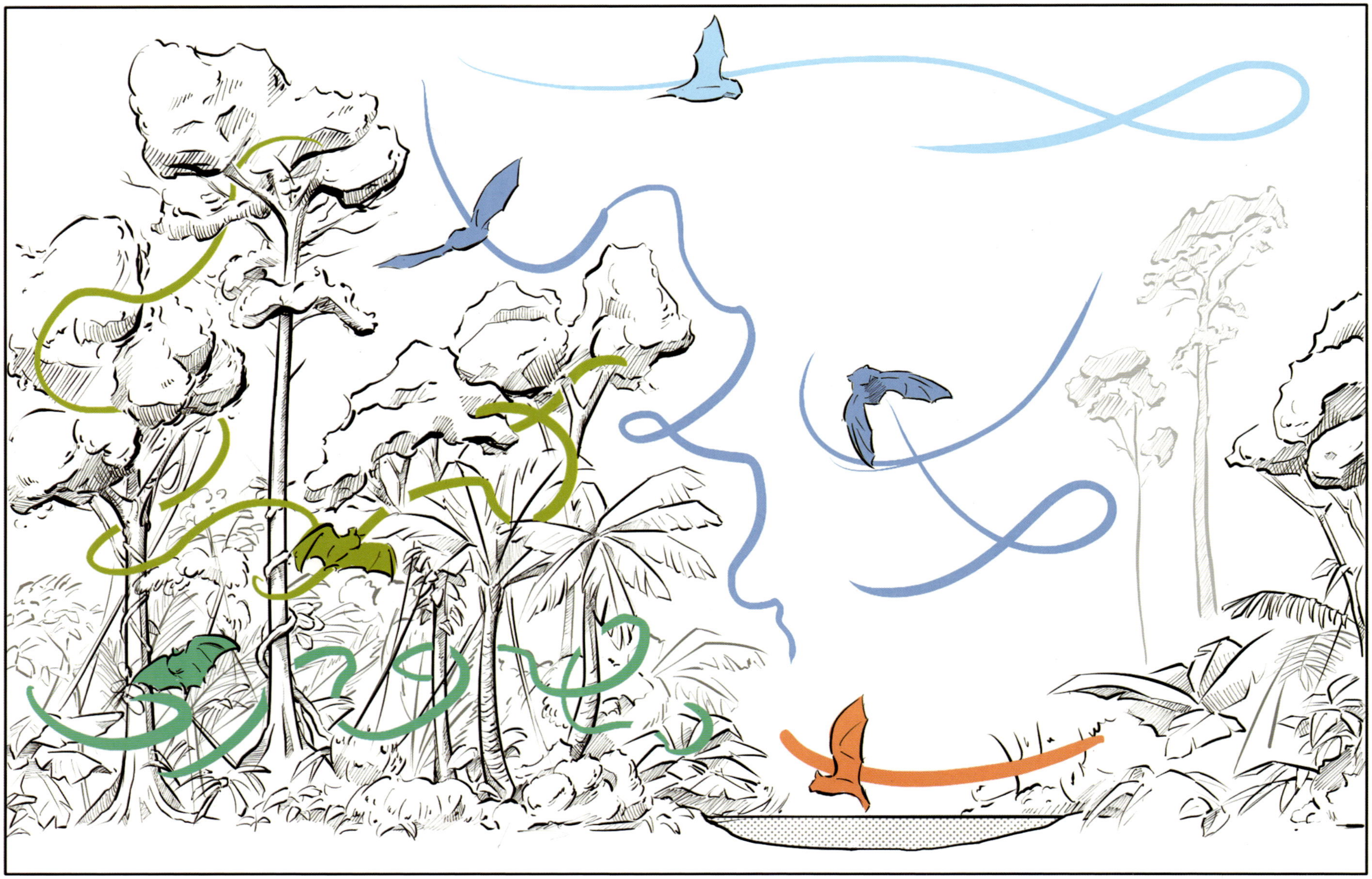

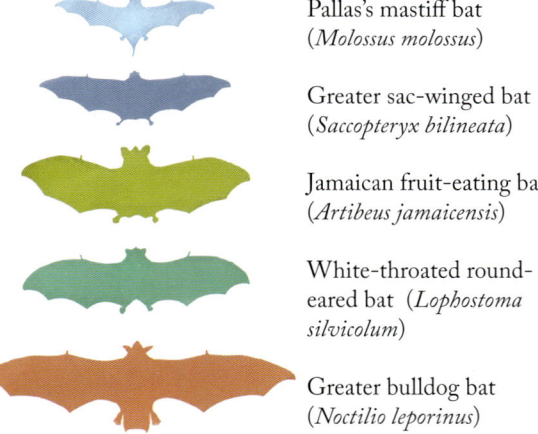

ABOVE A diversity of wing shapes allows bats to partition resources in the rainforest. Bats with long, narrow wings relative to their body size (Pallas's mastiff bat, *Molossus molossus*, light blue) forage in open spaces over gaps and above the forest canopy. Bats with wider, shorter wings relative to their body size forage closer to the forest edge (greater sac-winged bat, *Saccopteryx bilineata*, dark blue). The Jamaican fruit-eating bat (*Artibeus jamaicensis*, olive green) forages in the dense clutter of the forest canopy, while the white-throated round-eared bat (*Lophostoma silvicolum*, teal) is found in the understory. Both latter species have wide, short wings relative to their body size, allowing for high maneuverability. Others have more elongated, wide wings (for example, the greater bulldog bat, *Noctilio leporinus*, orange) that allow them to forage over water. Illustration © Javier Lázaro.

Pallas's mastiff bat
(*Molossus molossus*)

Greater sac-winged bat
(*Saccopteryx bilineata*)

Jamaican fruit-eating bat
(*Artibeus jamaicensis*)

White-throated round-eared bat (*Lophostoma silvicolum*)

Greater bulldog bat
(*Noctilio leporinus*)

VARIATION IN ECHOLOCATION CALLS

Wing shape as an indicator of foraging guild was not widely used in bat biology until it was combined with patterns discovered in echolocation calls. While there is tremendous diversity in individual echolocation calls, as described in chapter 3, general patterns emerge, with calls falling into broad categories that share acoustic characteristics such as frequency, bandwidth, and shape. These echolocation call categories correspond to specific foraging environments, such as open air, edges and gaps, dense vegetation, and water surfaces. By recording and analyzing the echolocation calls of a bat, these categories allow us to predict where a bat can forage without having to study it more closely. This realization was a breakthrough for the study of these elusive creatures because the echolocation call categories mapped onto corresponding wing shapes, reinforcing our understanding of the connection between a bat's foraging strategy and its flight environment. Bats with long, narrow wings adapted to foraging in the open air have louder, lower-frequency calls that work well for scanning over larger distances for insects, while bats foraging near the clutter of background vegetation have softer, more frequency-modulated, short calls that do not travel far, but give information about the surroundings at a higher resolution. Many more factors, such as physiological adaptations, interactions with other species, and roosting preferences then determine the structure of a species community at a given site.

Describing the ecological niche of a bat can be extremely difficult, as most bats are small and elusive. Even though the miniaturization and advancement of technology has taken huge steps in the recent past (see chapter 12), many bat species are too tiny and too fast for researchers to find their roosts, describe their diet, or understand their social systems. The realization that metrics such as wing shape and echolocation behavior could illuminate ecological niches and foraging guilds was a game changer. While previously bat researchers needed to catch a poorly known bat species,

Many aerial insectivores, like these chestnut sac-winged bats (*Cormura brevirostris*) shown here in their classic roosting position lined up one on top of another on a tree trunk, have distinctive echolocation calls, allowing them to be identified to species by call alone. Photo © Marco Tschapka.

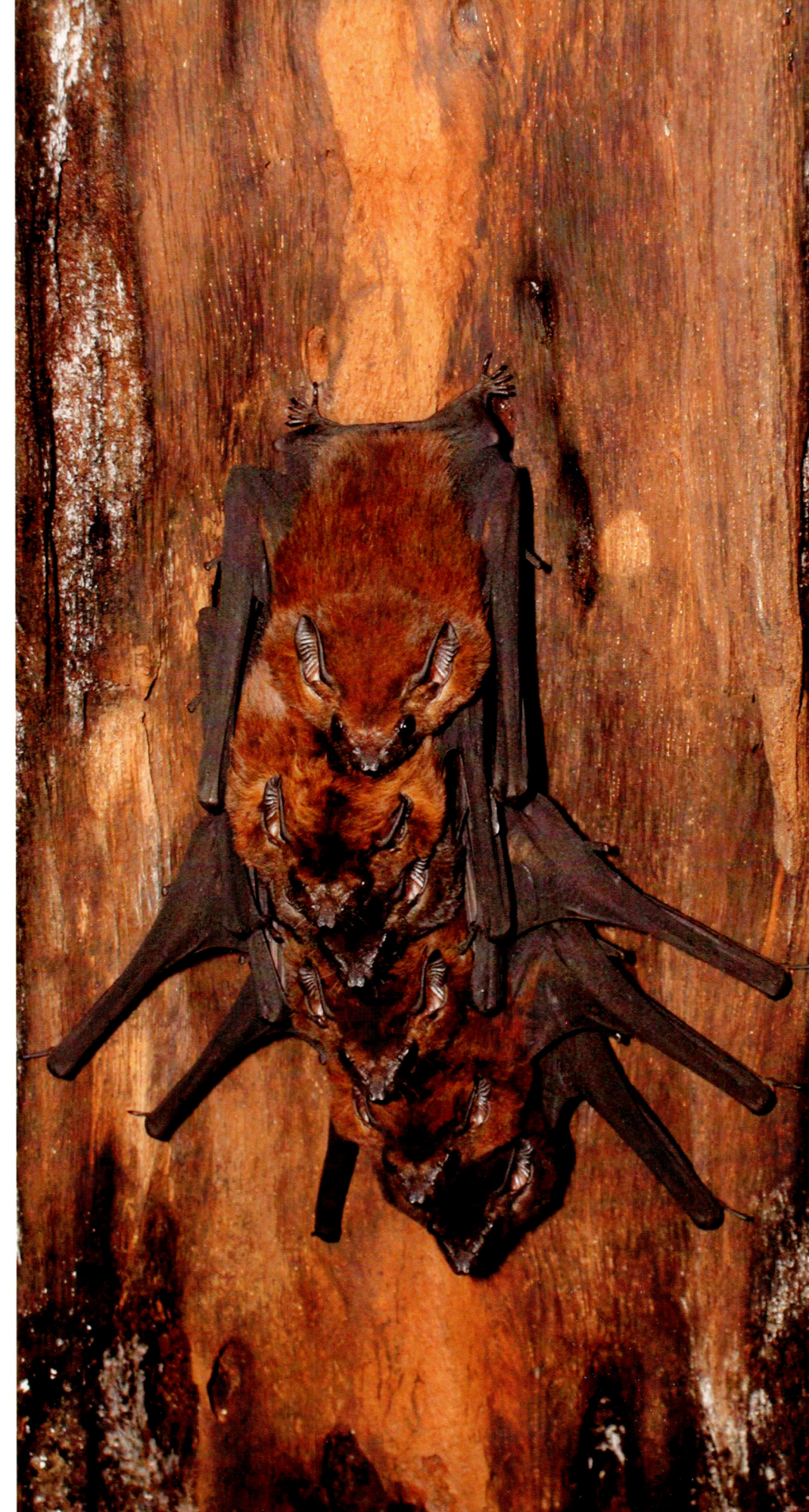

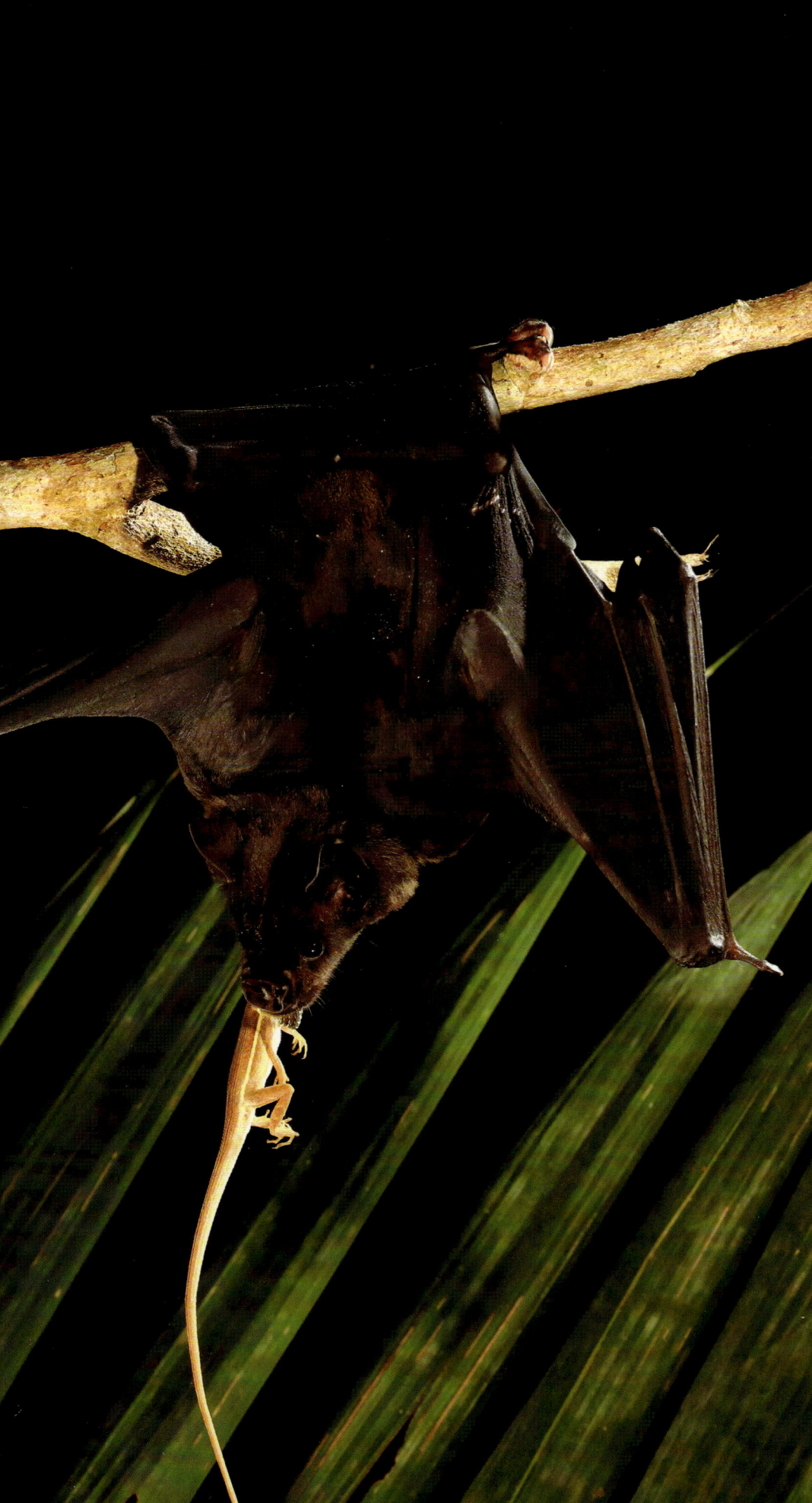

such as Sanborn's bonneted bat (*Eumops hansae*), in a mist net to obtain fecal samples to draw conclusions about its diet, researchers can now make predictions simply from a bat's wing shape or from its echolocation calls that determine if it hunts insects in the open air above the canopy in fast flight. Identifying these traits then allows statistical comparisons among species in a community to understand the partitioning of ecological niches, taking us a large step closer toward understanding how so many species of bats can coexist.

Further analysis can then link these traits to others to understand how suites of traits evolve and position species within a community. For example, by comparing wing shape to brain size, bat researchers can show that brains evolve for a particular ecological niche. Ancestral bats were of average wing and brain size, but as bats evolved, those that fly far in open spaces evolved smaller brains and longer, narrower wings, while bats that inhabit cluttered environments evolved larger brains and wider, shorter wings. When bat species entered a niche that demanded large energetic trade-offs in a simple flight environment, a large brain was too costly. The narrow-winged *E. hansae*, for example, forages in open spaces and has a smaller brain than its ancestors. A large brain would be costly for this streamlined fast flyer, and also unnecessary. The open-air environments it navigates are much less complex than the forest understory, where a bat such as the white-throated round-eared bat (*Lophostoma silvicolum*) forages for insects sitting on leaves. This species has a large brain and broad, short wings to maneuver through the complex forest understory.

The understanding gained from the guild concept can be particularly important in today's changing world, where many poorly known species are facing habitat loss and human encroachment, and time and funding for in-depth studies are not available. Knowing which species are more flexible and able to adapt to environmental changes and which need rapid conservation attention is increasingly critical for their survival.

LEFT Greater spear-nosed bats (*Phyllostomus hastatus*) are some of the most formidable predators of the bat world and will feed on insects, lizards, and even on other bats. They are omnivores that make full use of what the forest has to offer, eating fruit and nectar as well. OPPOSITE A Jamaican fruit-eating bat (*Artibeus jamaicensis*) plucking a ripe fig.

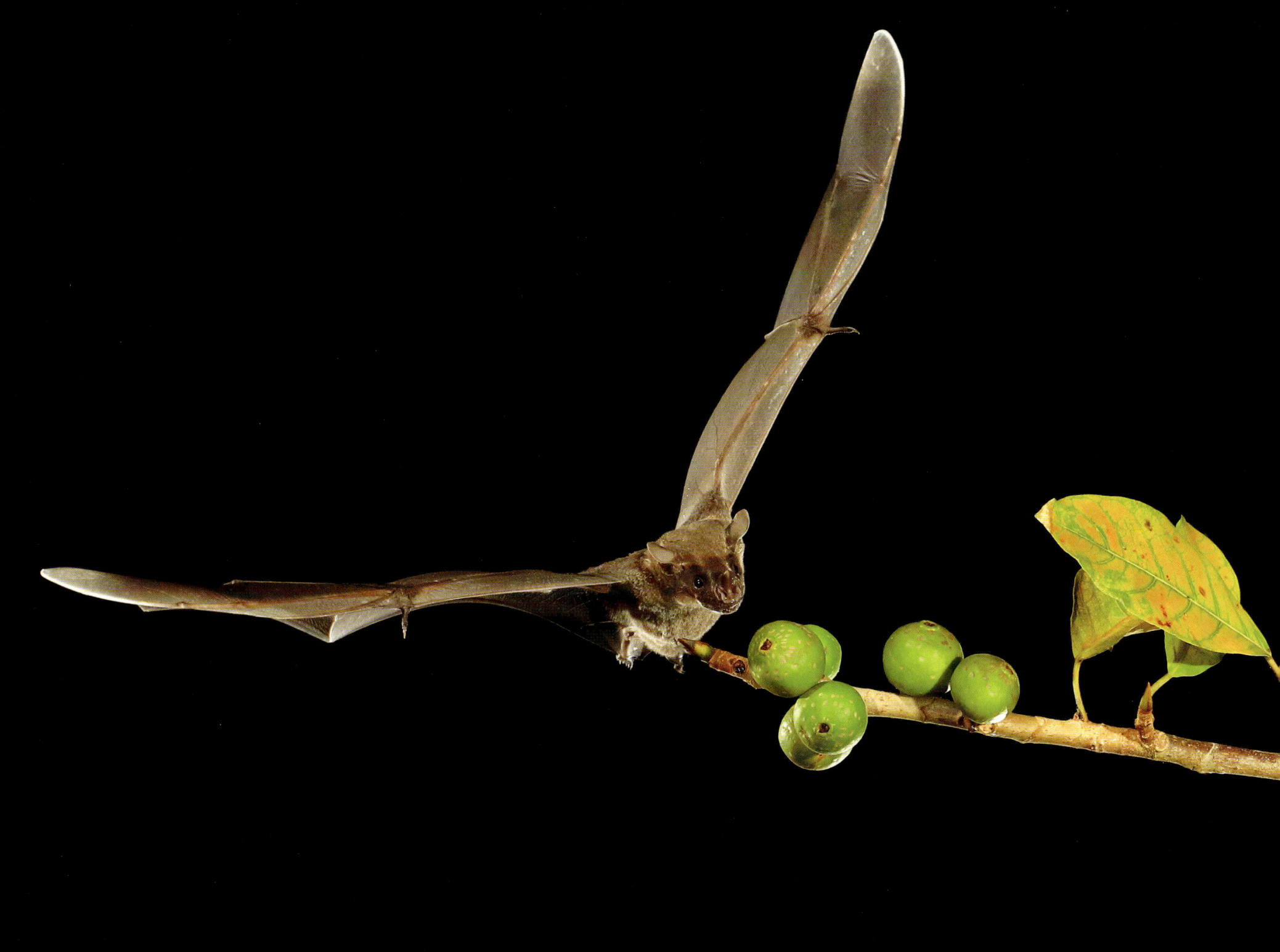

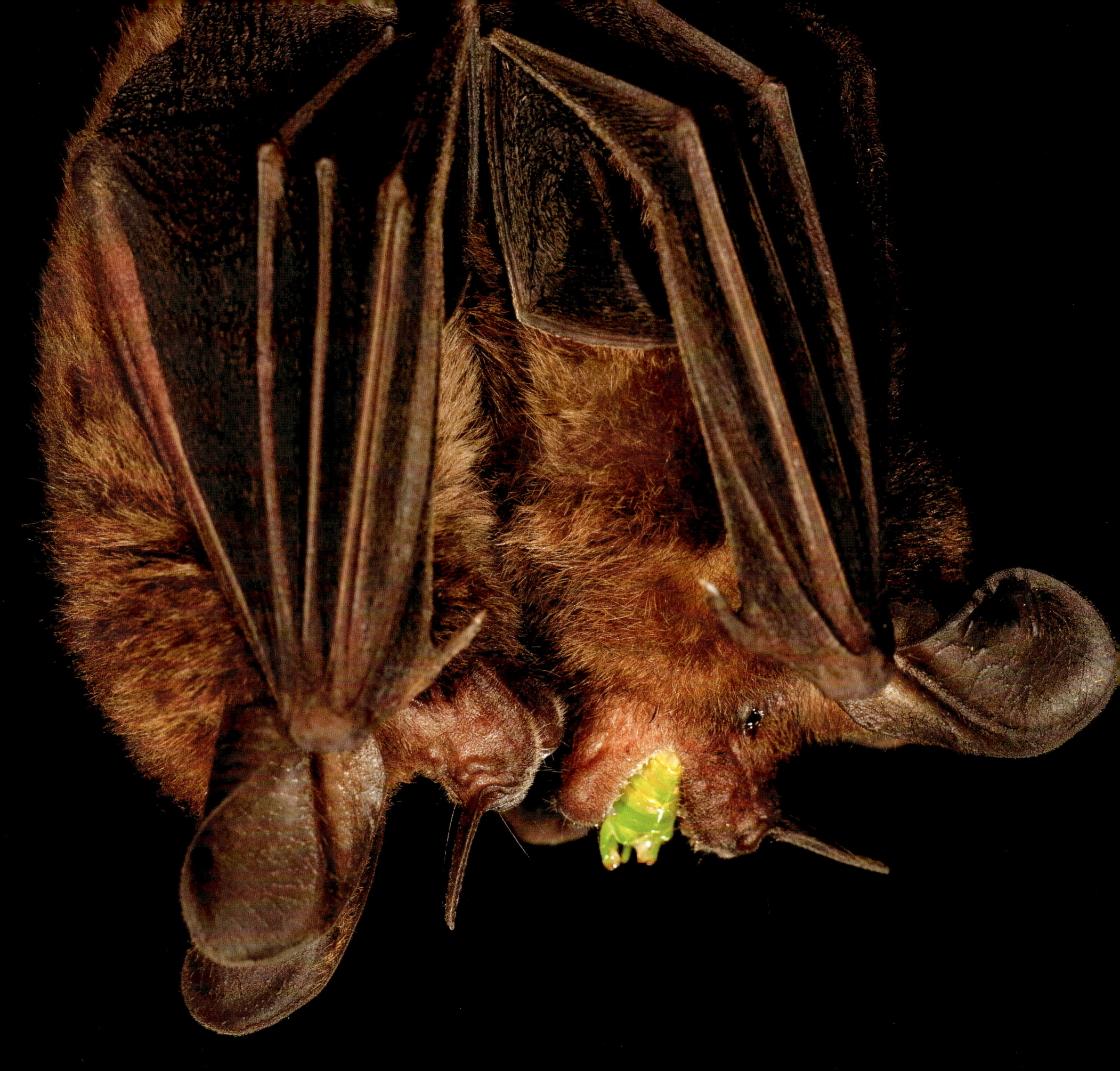

CHAPTER SIX

DINNER IS SERVED

GOURMETS AND GOURMANDS

Bats from the Neotropics feed on the widest range of foods of any group of mammals. Fruit, pollen, nectar, small vertebrates like birds, fish, and even other bats make up the menu of the bat community on BCI. Many tropical bat species feed on small flying insects, like their cousins from the temperate zones, but tropical bats show dramatic adaptations that allow them to feed on a diverse range of foods. Omnivorous bats like the greater spear-nosed bat (*Phyllostomus hastatus*) will eat nearly anything smaller than them—fruit, insects, lizards, and occasionally even

smaller bats. With a wingspan close to 3 feet (1 meter), the spectral bat (*Vampyrum spectrum*) is the largest bat in the Americas and lives mostly on vertebrate prey—birds and rodents they capture in the dark. Bone middens under the roosts of this large predator reveal the hidden vertebrate diversity of the forest.

Despite the dietary diversity, for nearly eighty bat species to coexist on a small island like BCI, they need to split resources in a way that allows them to minimize where, when, and how they compete. Some bats are gourmets, feeding on a select and often

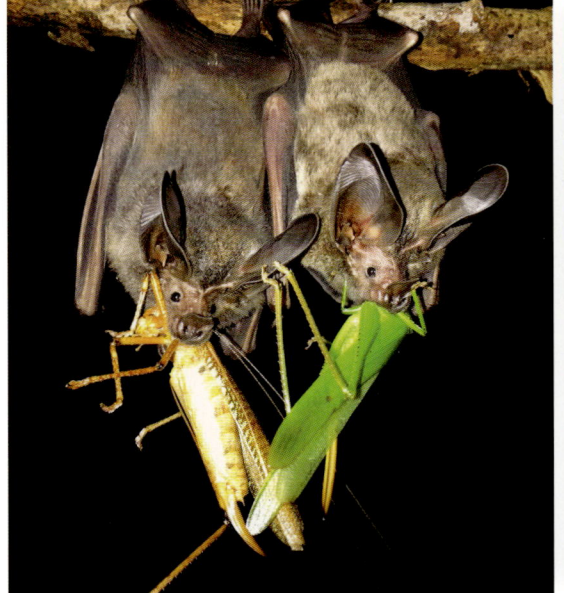 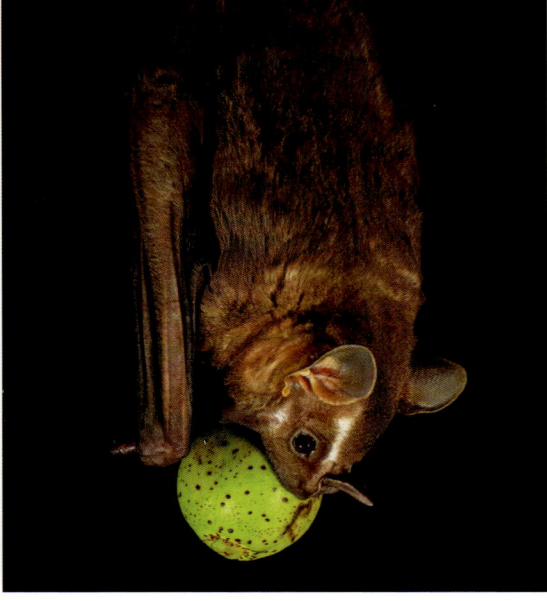 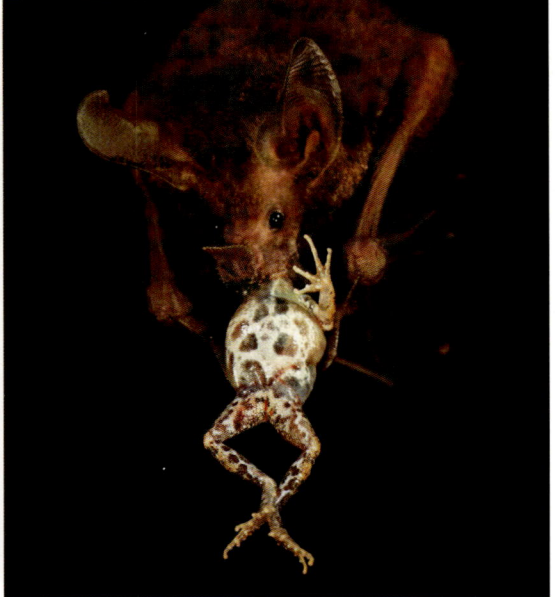

OPPOSITE A pair of common big-eared bats (*Micronycteris microtis*) cuddle belly-to-belly, while one finishes the last few bites of a caterpillar. **ABOVE LEFT TO RIGHT** White-throated round-eared bats (*Lophostoma silvicolum*) consume katydid prey. A great fruit-eating bat (*Artibeus lituratus*) eats a tasty fig. A fringe-lipped bat (*Trachops cirrhosus*) consumes one of its favorite frogs (*Engystomops pustulosus*).

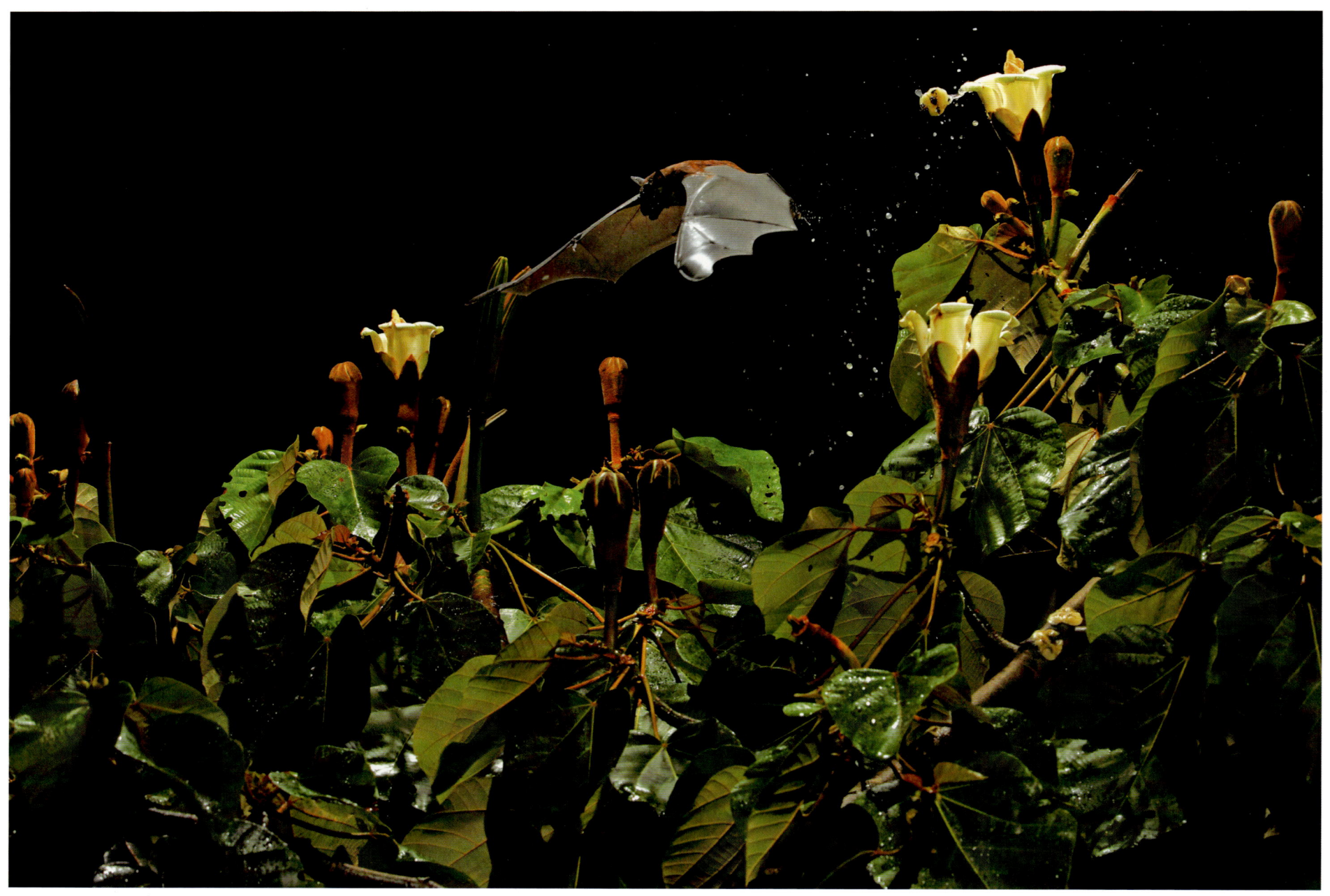

A greater spear-nosed bat (*Phyllostomus hastatus*) takes off after landing briefly on a balsa flower (*Ochroma pyramidale*) to drink the nectar. Balsas tend to flower when no other fruits or flowers are available, and greater spear-nosed bats will travel long distances to find them.

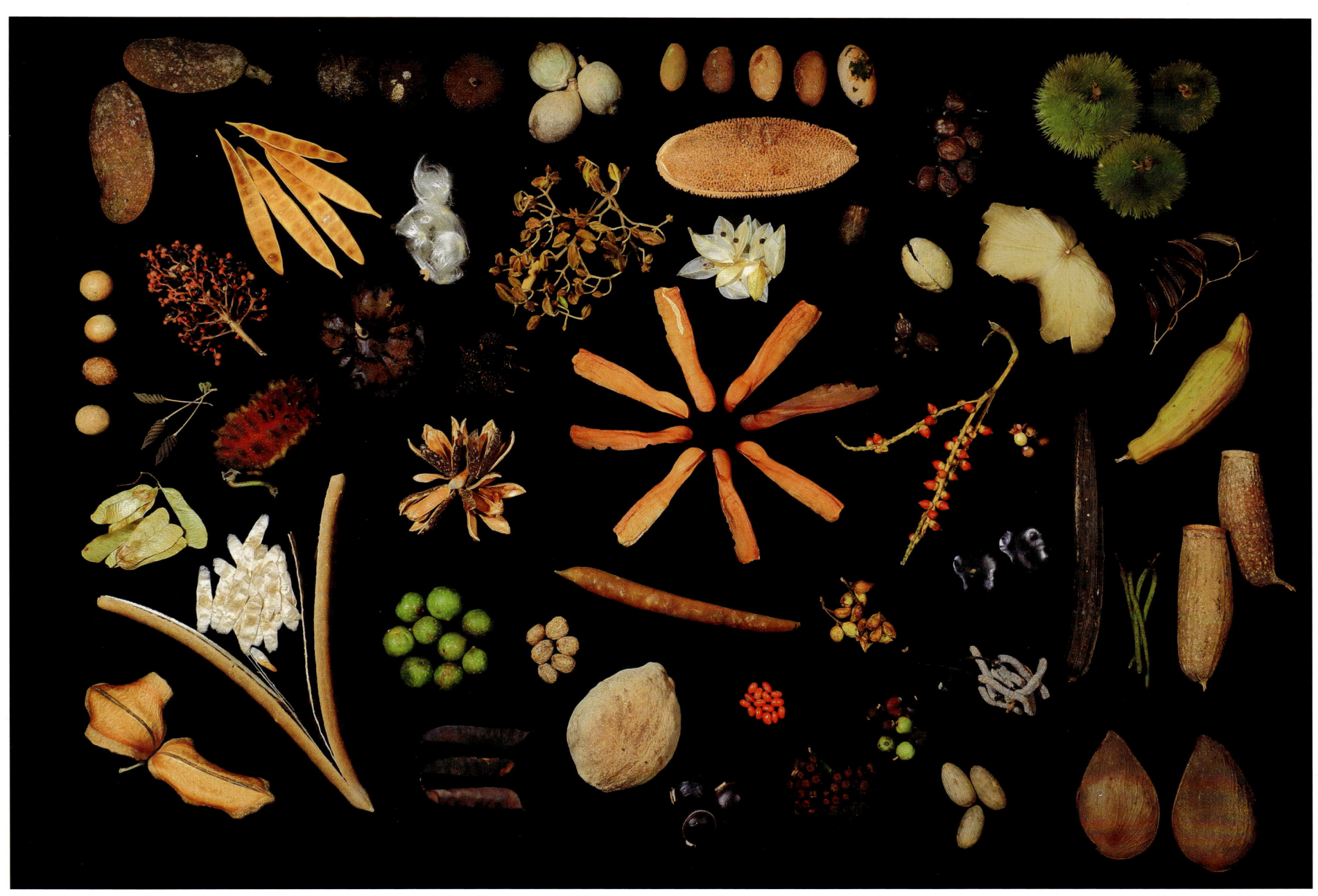

Barro Colorado Island hosts a great diversity of fruits and seeds. While some are transported by the wind to a new home, many rely on the transport services of bats and other frugivorous animals.

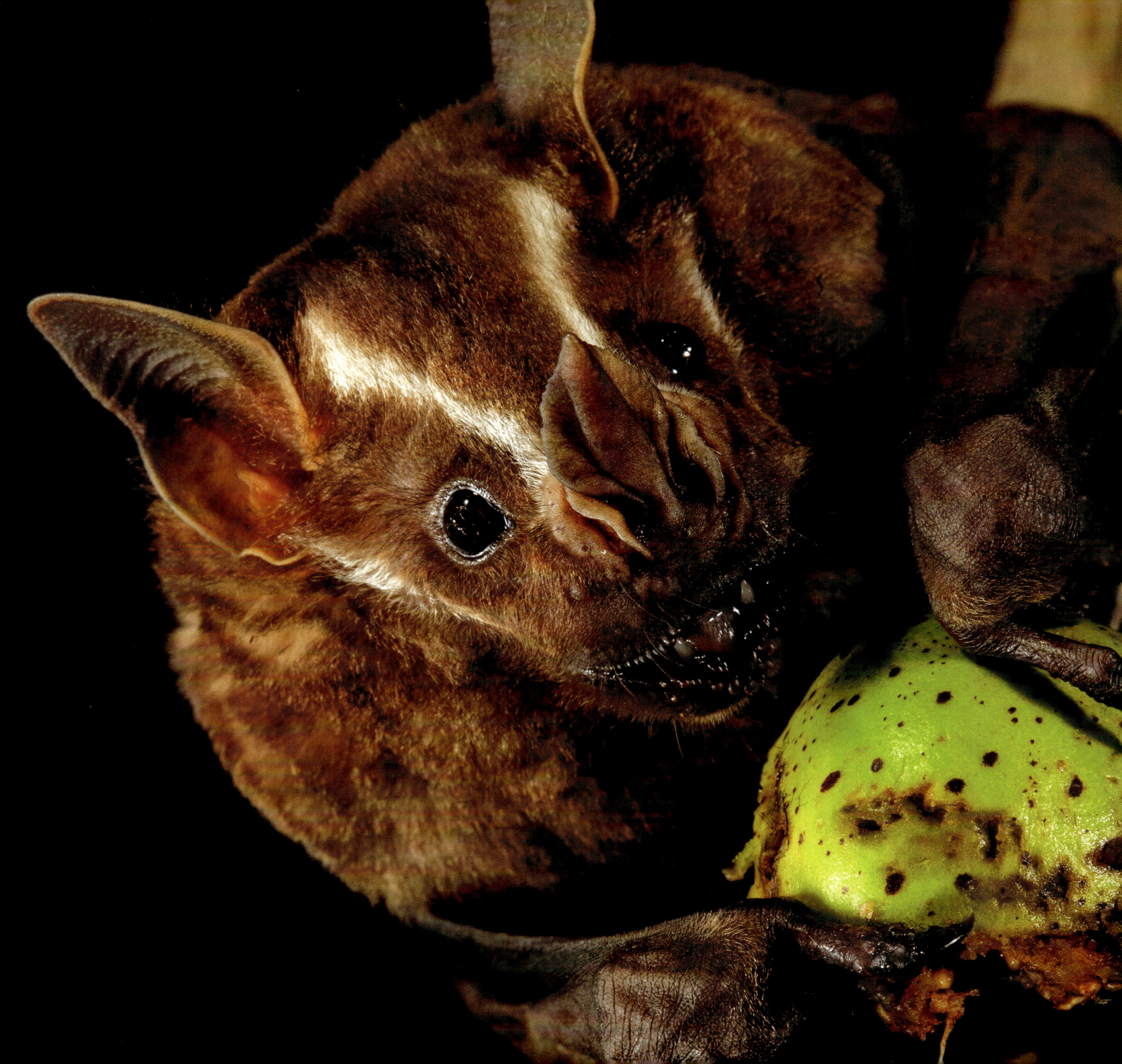

quite specialized menu, while others are gourmands and gather their food from a wide menu. The differences in dietary makeup are one part of niche partitioning, an evolutionary strategy that allows many seemingly similar species to coexist in the same space (see chapter 5). For bats, much of this niche partitioning is structured by body size. For example, among the large group of bats that feed mostly on figs, small bats prefer smaller-sized fig species and larger bats prefer larger fig species. Bats can recognize appropriately sized figs by their smell alone. When offered ripe figs hidden in cloth bags, fig-eating bat species choose figs of a matching size class. It's not that small bats cannot eat large figs—they will routinely carry off figs that weigh as much as 60 percent of their body mass—but given a choice, they find the smaller fig species more attractive. Many fig trees in the tropics rely on bats to disperse their seeds. To attract frugivorous bats, fig trees perfume their green fruit and display it conspicuously to facilitate bat detection. In contrast, figs dispersed by birds are usually red and do not have a strong smell (see chapter 9). Even biologists can smell the species-specific bouquets of these ripe fruits when standing under a fruiting tree.

A similar size correspondence is found in the short-tailed bats (*Carollia* spp.). These bats are well-known for feeding on the fruit of tropical pepper (*Piper* spp.) plants, sometimes known as *candela* in Latin America because their greenish-white fruits stand up straight like a candle. Only a few fruits on each pepper shrub ripen each night. This forces the short-tailed bats to travel far to find enough fruits to eat. The small chestnut short-tailed bat (*C. castanea*) feeds exclusively on pepper plants and monitors all of the plants in its home range, while the larger Seba's short-tailed bat (*C. perspicillata*) will feed from a wider diversity of fruits—even bananas and papayas left in feeders for tropical birds. This balance of smaller gourmets and larger gourmands contributes to keeping competition to a minimum.

Other gourmands take full advantage of the tropical buffet. The common big-eared bat (*Micronycteris microtis*), for example, can find insects that sit quietly and motionless on a leaf with their fine-tuned echolocation. Their stealth hunting technique allows them to find a wide variety of prey, including caterpillars, dragonflies, and stick insects—all foods that are very unusual for a bat. These tiny, yet fierce hunters will also capture and feed on hairy caterpillars and armored katydids that are larger than they are. As hunters of anole lizards, they are among the smallest predators of vertebrates in the world.

Perhaps the most extreme examples of a gourmet are vampire bats. The three species of vampire bats, which are only found in the Neotropics, feed exclusively on blood but target different animals to harvest it from. Common vampire bats (*Desmodus rotundus*) seek out mammals, while the white-winged vampire bat (*Diaemus youngii*) and the hairy-legged vampire bat (*Diphylla ecaudata*) prefer birds. These bats are extreme specialists and have incredible adaptations to allow them to feed on blood. They are the most terrestrial (ground-based) of bats in the Neotropics, and can run and jump on the ground. They have heat-sensors in their muzzles, their kidneys are specialized to deal with high-water-content food, and they have some of the most intricate social relationships in bats. Since it can be difficult to find a sleeping animal to feed from in a natural forest, individuals returning hungry to the roost receive regurgitated blood from groupmates. On more successful days they will then return this favor.

The community of gourmets and gourmands allows for high bat diversity in many tropical locations, including BCI. An intricate partitioning of resources, with some species highly specialized while others sample from a broad menu, is one of the ways that tropical forests support such a large species diversity. The rapid diversification and radiation of the leaf-nosed bats, for example, started with a generalist gourmand ancestor with the evolutionary and ecological flexibility for many species to subsequently develop feeding ecologies specialized on particular food types, including insects, vertebrates, blood, fruits, and flowers. The balance of gourmets and gourmands drives bat diversity, with these feeding groups characterizing the way we think about many of the species in the tropics.

OPPOSITE Bats can adeptly manipulate food using just their thumbs. A great fruit-eating bat (*Artibeus lituratus*) feeds on a fig. PAGES 80–81 A greater spear-nosed bat (*Phyllostomus hastatus*) feeds on the nectar pooled in balsa flowers (*Ochroma pyramidale*) by landing briefly to drink, dusting its face and abdomen with pollen in the process.

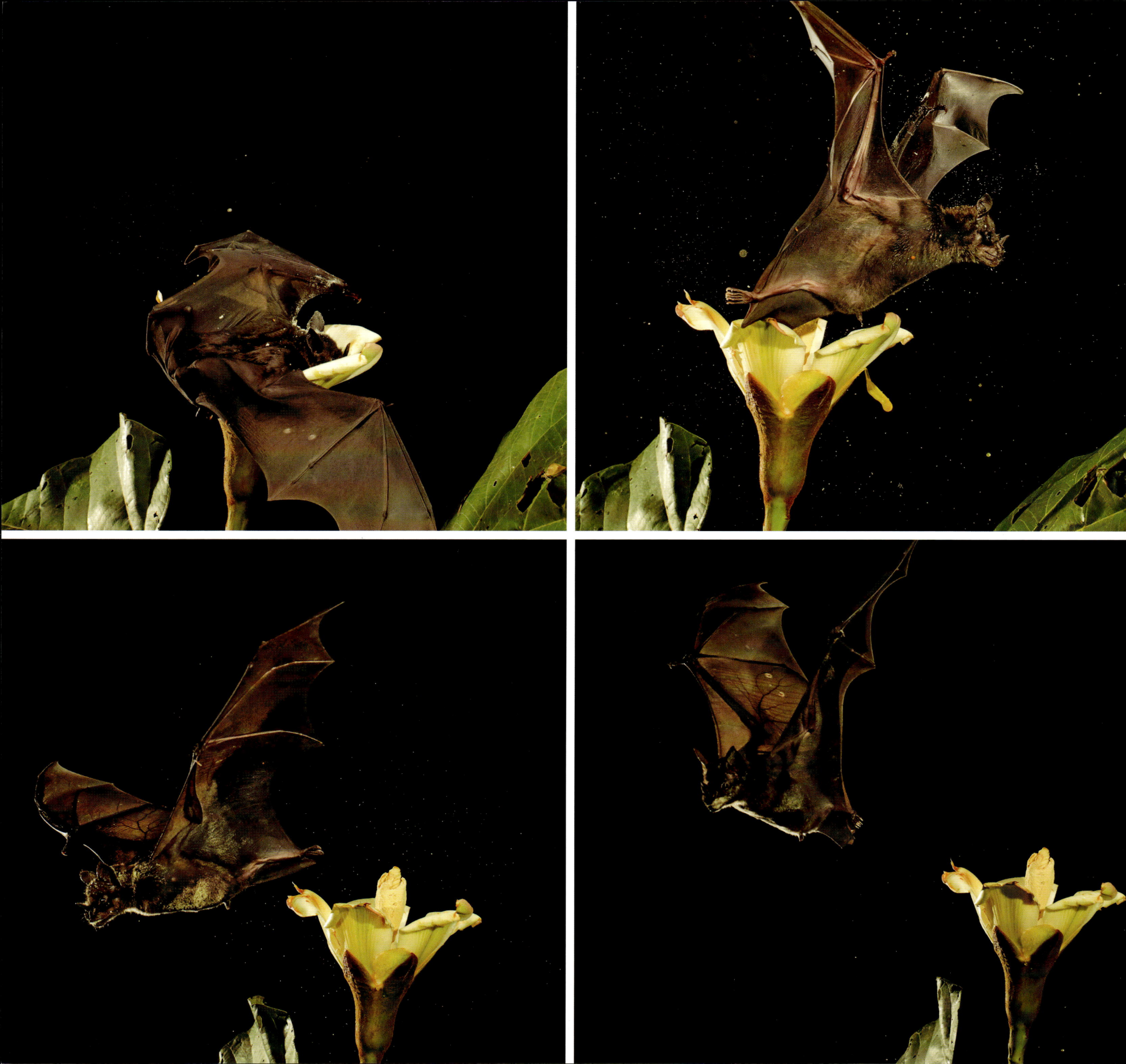

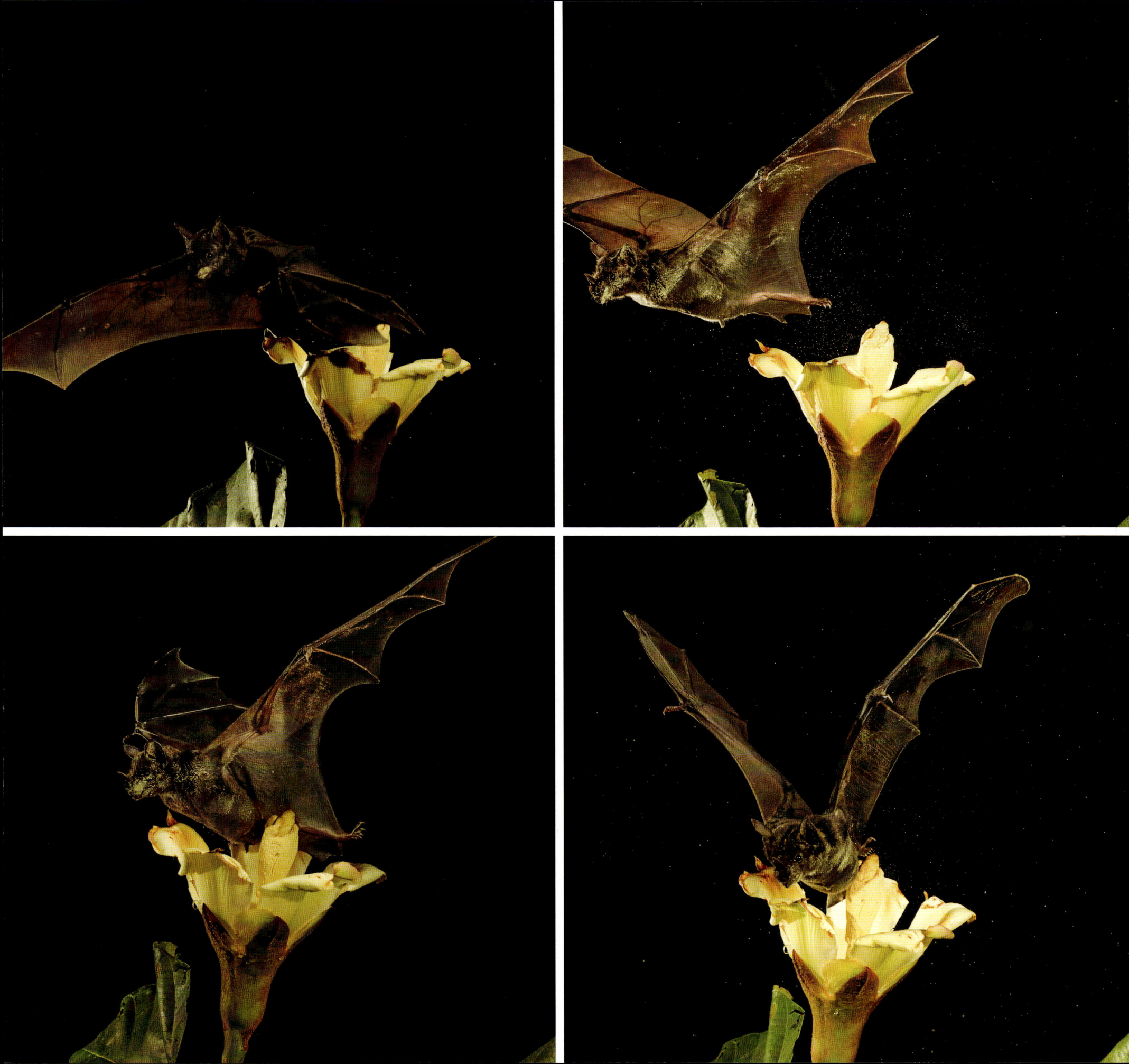

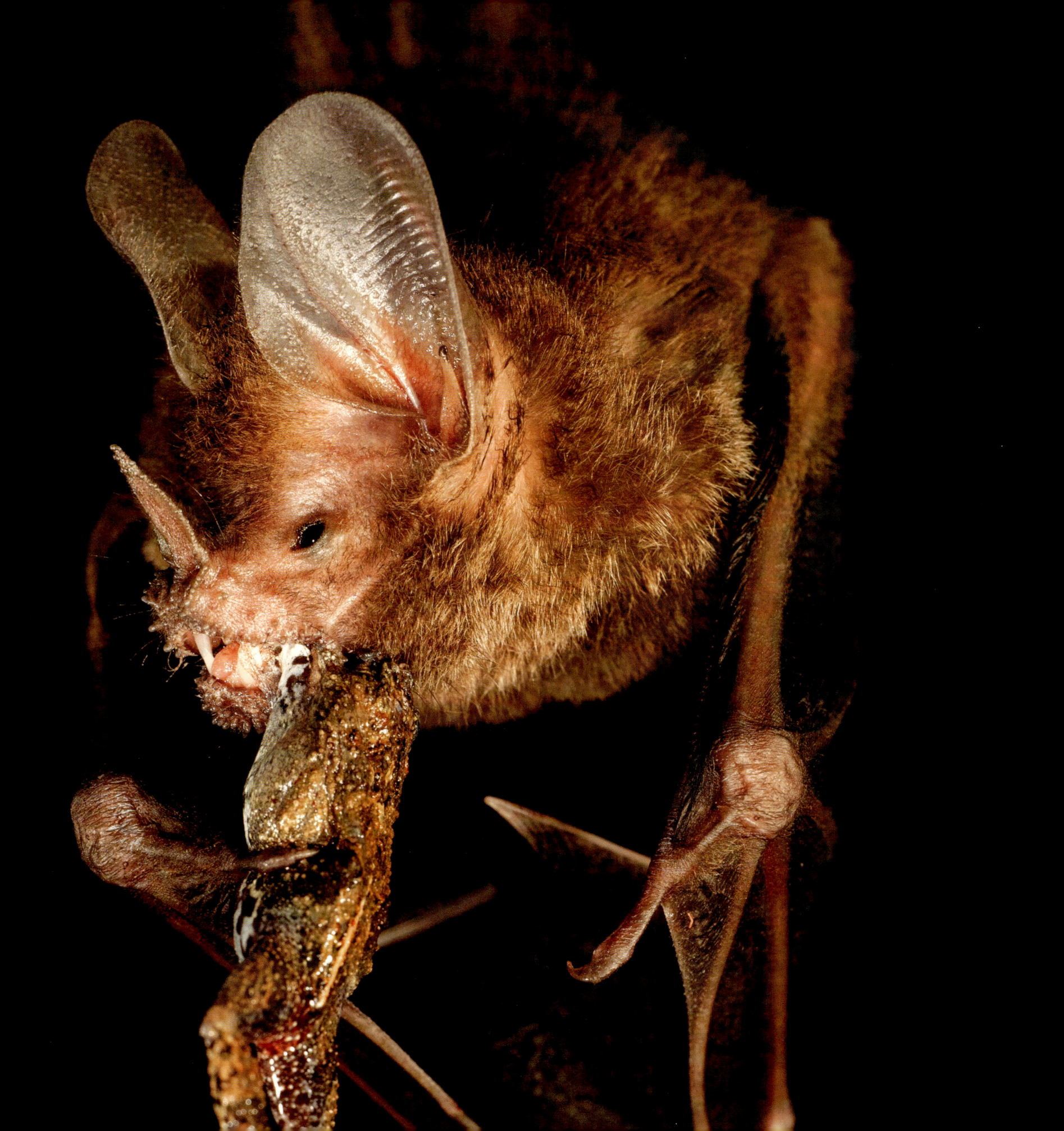

CHAPTER SEVEN

OF HUNTERS AND FISHERS

FORAGING STRATEGIES OF ANIMAL-EATING BATS

Among animal-eating bats, species have evolved a broad range of feeding strategies. Most feed on insects, but prey can also include fish, birds, rodents, frogs, lizards, and even smaller species of bats. A corresponding myriad of hunting strategies have evolved, with each species adapted to its distinct foraging niche. The most common foraging strategy among bats is to catch insects in the air, but they do this in many different ways, reflected in their specific morphologies and echolocation behaviors. These distinct foraging strategies are major drivers of the diversity of bat species on BCI and worldwide. As we saw in previous chapters, some insectivorous bats hunt swiftly far above the canopy, while others search in a slower and more convoluted way along the edges of vegetation and in forest gaps. The effects that insectivorous bats have on their

prey are manifold. Some groups of moths have evolved ultrasound hearing, enabling them to detect and evade hunting bats. Bats in turn have evolved more sophisticated echolocation strategies, in an ever-escalating arms race between predator and prey.

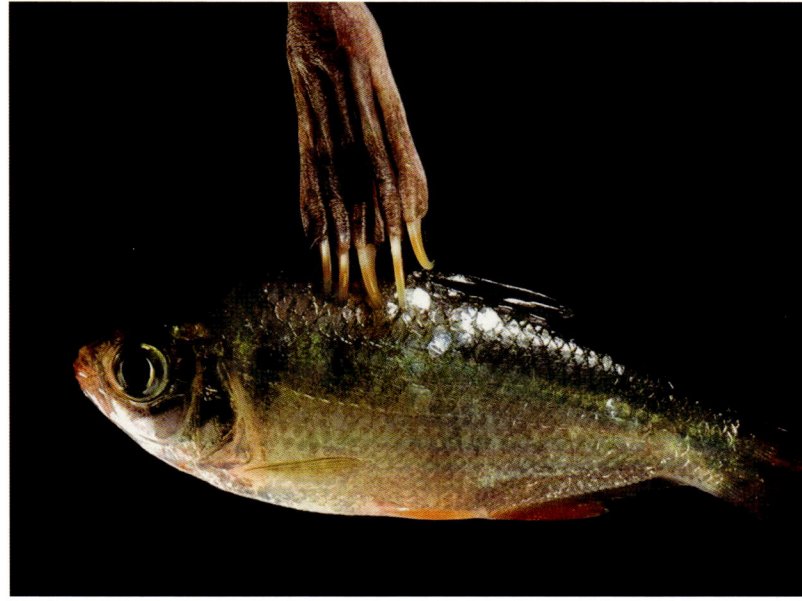

OPPOSITE After capturing the frog from a pool, a fringe-lipped bat (*Trachops cirrhosus*) carried this túngara frog (*Engystomops pustulosus*) back to a perch to feed. ABOVE The long and sharp claws of the fishing bat are perfect to hold even slippery fish firmly.

Some insectivorous bats eat so many insects that they are considered effective pest control agents, saving farmers billions in damaged crops and pesticide use each year. While both insectivorous birds and bats help to protect plants from being consumed by insects, bats are particularly effective insect predators. Exclosure experiments on BCI showed that when insectivorous birds were not able to access plants and the insects on them, the plants suffered a 67-percent higher herbivory damage rate compared to controls. However, when bats were denied access, herbivory damage rates rose 209 percent compared to controls, highlighting the importance of bats in curbing insect damage to plants.

Rather than capturing flying prey, gleaning bats take food from surfaces, often in dense rainforest understory. Because echoes bounce off the complex surfaces of vegetation in many different directions, distinguishing prey from background clutter by echolocation is challenging for bats in these environments. For this reason, while gleaning bats use echolocation to navigate, many find their prey instead by listening for prey-produced sounds, such as prey locomotion noises or prey mating calls. For example, male túngara frogs (*Engystomops pustulosus*) gather in choruses to attract females. The fringe-lipped bat (*Trachops cirrhosus*) eavesdrops on these advertisement calls and uses them to find its prey. *T. cirrhosus* is just one example of many leaf-nosed bats that have sensory and morphological adaptations which allow them to locate their prey by attending to prey-emitted sounds.

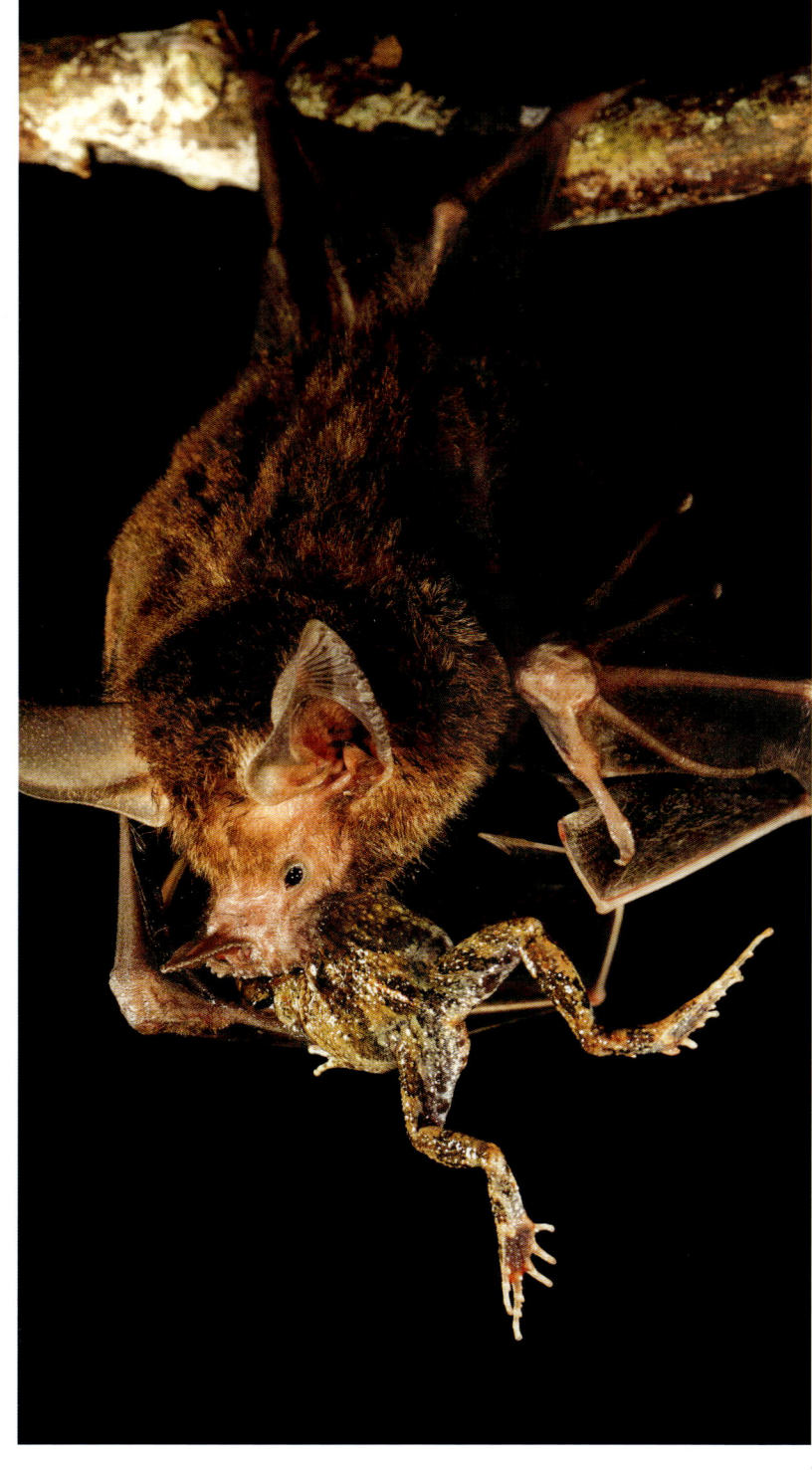

A fringe-lipped bat (*T. cirrhosus*) returns to its hunting perch to eat a túngara frog (*E. pustulosus*).

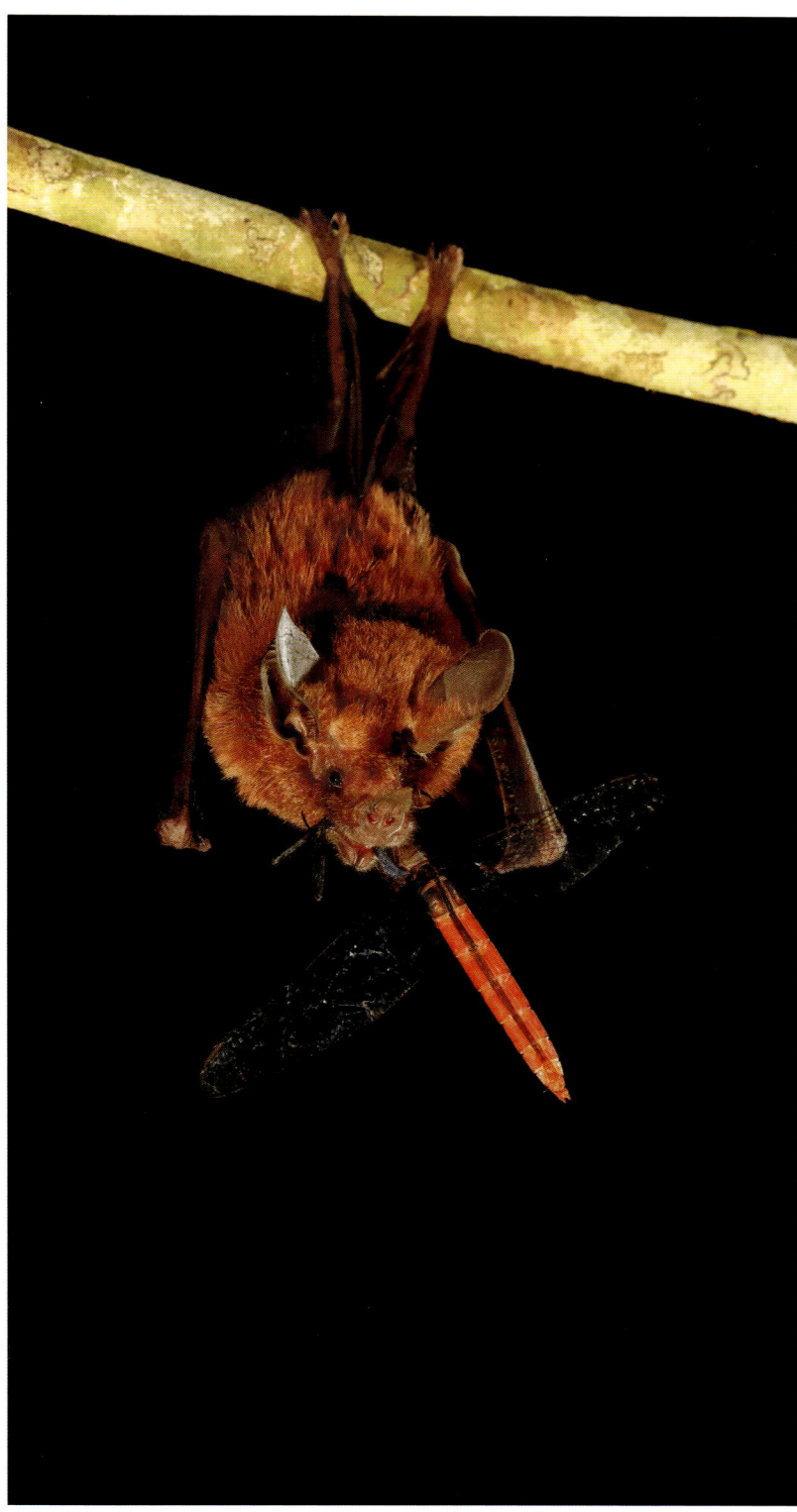

It was long thought that gleaning silent, motionless prey in dense forest understory was physically impossible using echolocation due to the masking echoes already discussed. However, initial observations of the common big-eared bat (*Micronycteris microtis*) on BCI showed that this bat feeds heavily on dragonflies. Dragonflies are exclusively diurnal insects that spend their nights silently sleeping motionless on branches. Yet *M. microtis* finds them. Over the last years, detailed experiments on BCI have shown that *M. microtis* uses optimal approach angles and sophisticated echolocation to systematically scan the vegetation for a variety of sleeping prey, from butterflies to anole lizards.

A common big-eared bat (*Micronycteris microtis*) beginning to consume a recently captured dragonfly. Sophisticated echolocation allows *M. microtis* to detect silent, motionless insects, even those asleep on forest vegetation at night.

In a special case of gleaning, greater bulldog bats (*Noctilio leporinus*) trawl water surfaces with their huge, laterally compressed feet to hunt for small fish. The fish are safe as long as they are under water, because echolocation calls bounce off the water surface. So, *N. leporinus* captures fish in two different ways: They use echolocation to detect small ripples where a fish has broken the water surface and then directly target this location by dipping their large claws into the water. Or they rake their large feet over several feet (meters) along the water surface in patches with many surfacing fish in what is called a "directed random rake." When they spear a fish with their large claws, they transfer the fish in flight from their feet to their huge cheek pouches, where they store it while they continue to hunt. Because fish often occur in swarms, *N. leporinus* often hunts in groups. Where there is food for one, there is food for many, and they can share.

Another, much smaller BCI bat, the long-legged bat (*Marcrophyllum macrophyllum*) uses a similar hunting strategy to target insect prey at and above the water's surface. This is the only bat from the Phyllostomidae family that, in a remarkable evolution, gained access to aquatic prey by adopting a trawling strategy.

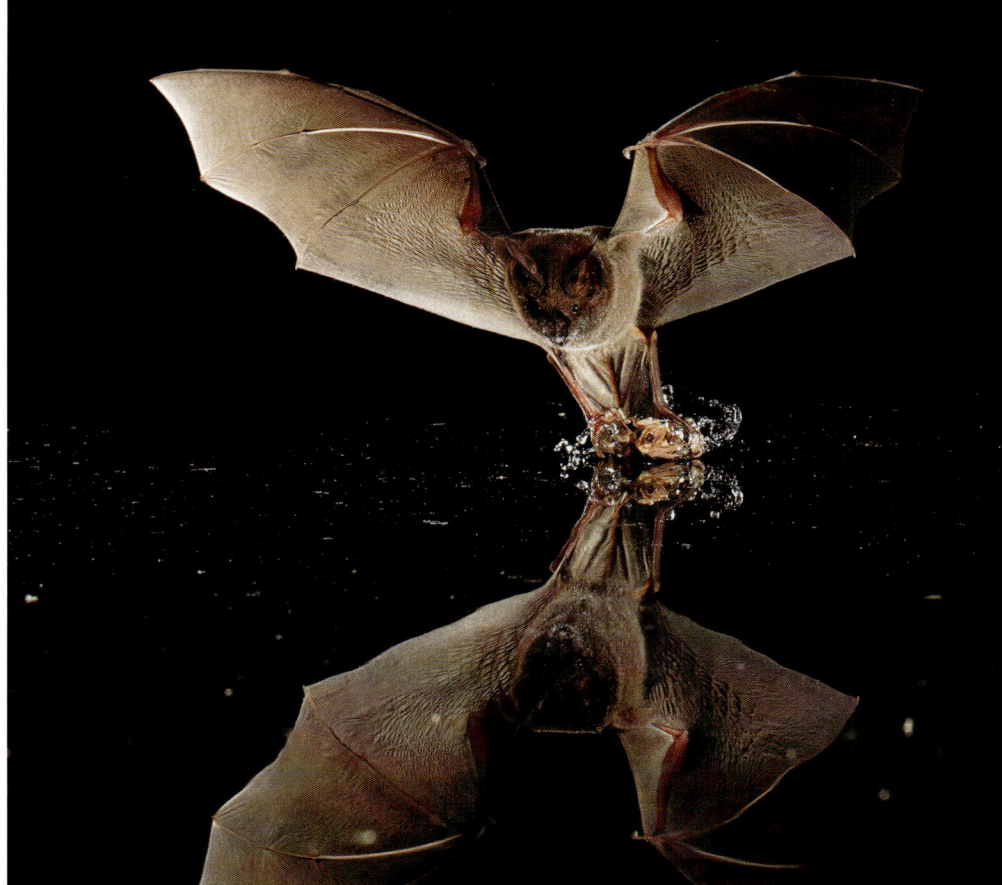

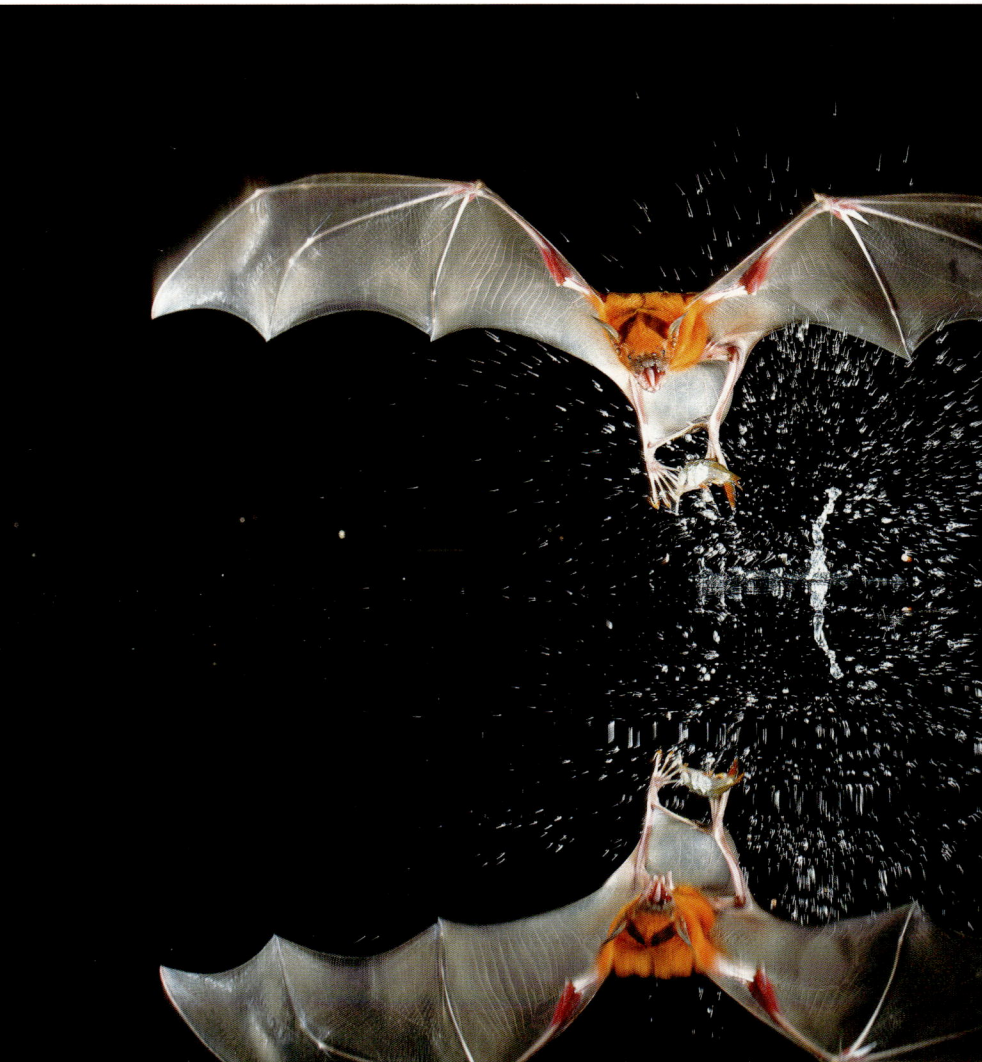

Trawling the water's surface for food is a rare behavior but has evolved in several different bat lineages across the world. TOP Long-legged bats (*Macrophyllum macrophyllum*) glean insects off of the water's surface. BOTTOM Greater bulldog bats (*Noctilio leporinus*) rake the water's surface for surfacing fish.

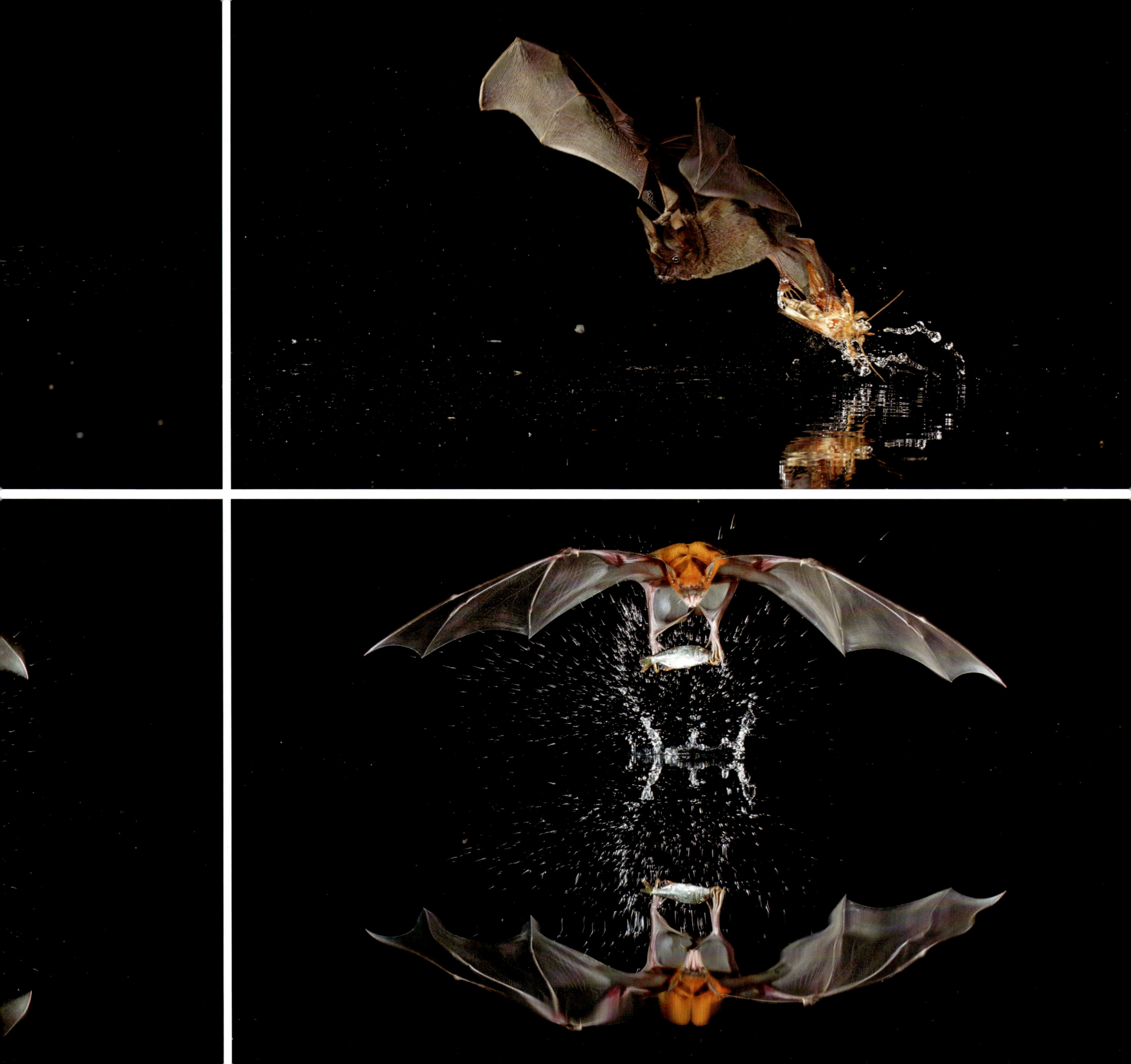

RIGHT A spectral bat (*Vampyrum spectrum*) finishes its meal of a smaller bat species, leaving only the wing behind. Photo © Marco Tschapka.
OPPOSITE A greater bulldog bat (*Noctilio leporinus*) feeds on a fish and shows off its enlarged cheeks. Greater bulldog bats have elastic cheeks and specialized cheek pouches that allow them to store captured fish while flying over large bodies of water.

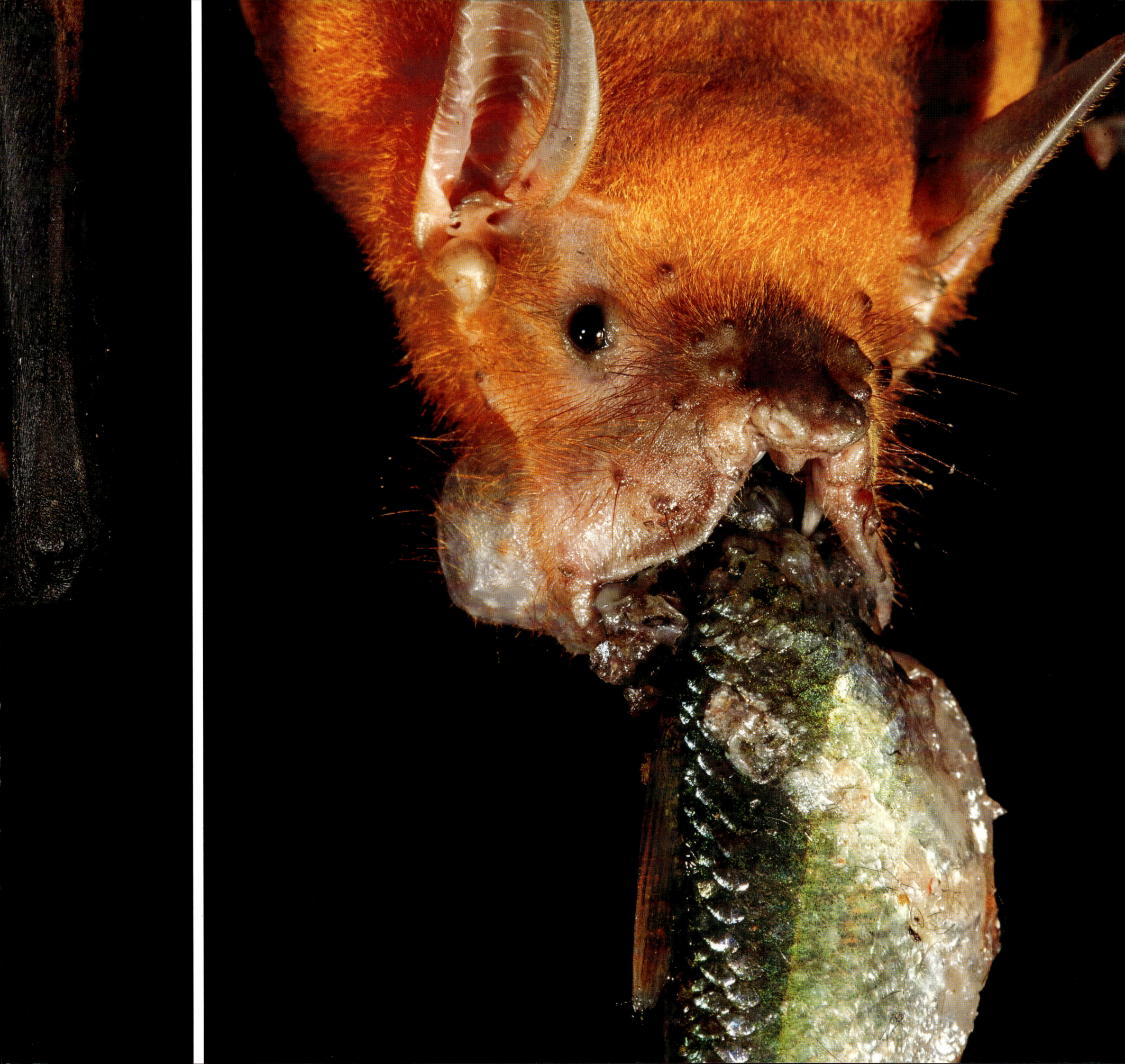

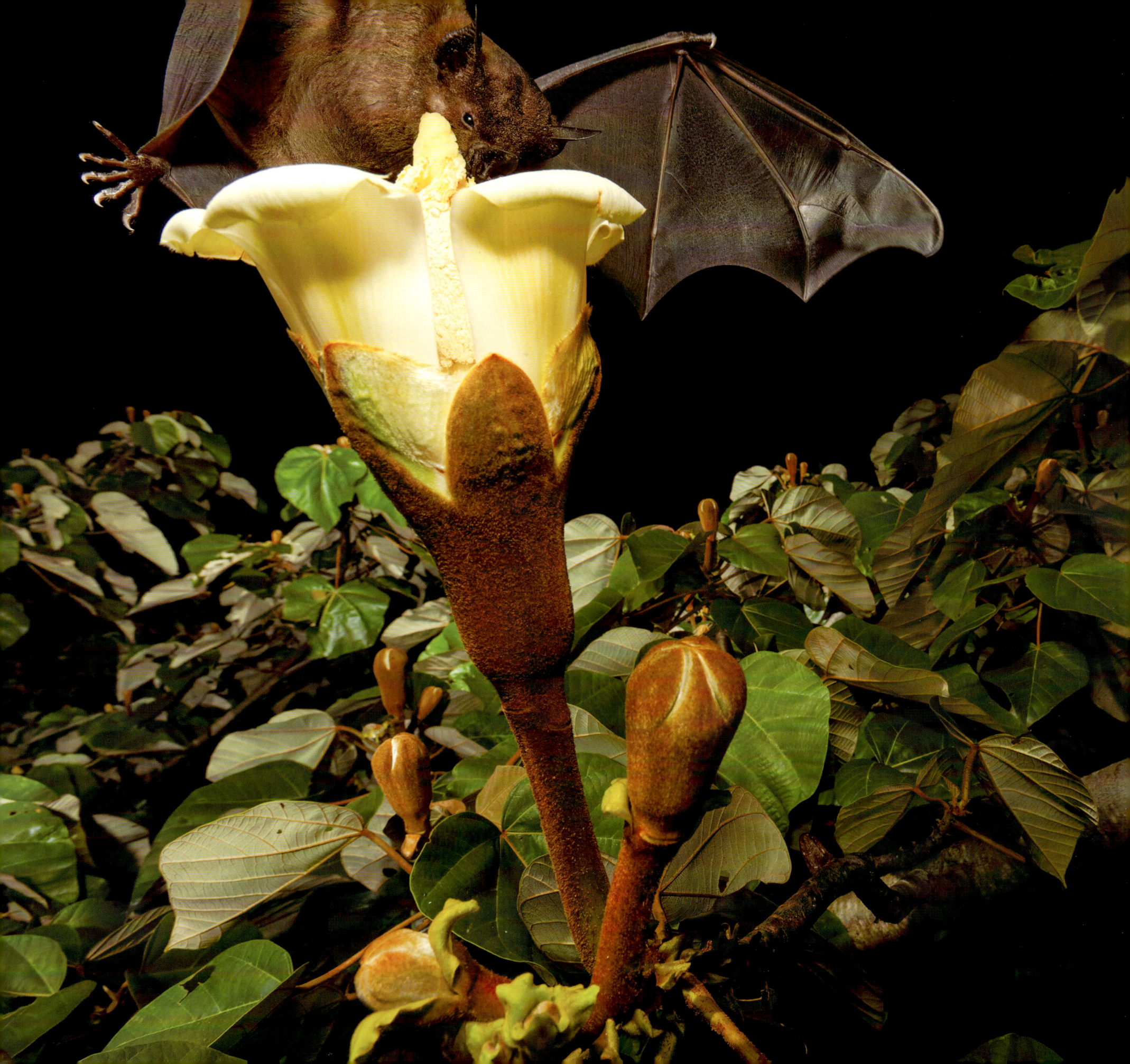

CHAPTER EIGHT

NOCTURNAL POLLINATORS

BAT FLOWERS AND FLOWER BATS

In temperate zones, all bat species feed on insects, but the tropics offer a number of alternative foraging options. Only in the tropics will you find vegetarian bats that lick nectar from flowers or that live partly, or entirely, on a fruit diet. At first glance, feeding on fruit or nectar does not seem so different from feeding on insects. However, for a foraging bat these two strategies pose completely different challenges. Insectivorous bats have to search for prey that tries its best to evade detection. In contrast, many plant species depend on bats for pollination and seed dispersal, and thus facilitate this task by advertising their goods through various sensory channels.

Plants lack one critical ability: they cannot move. This is problematic during reproduction, one of the most critical moments in any life. To produce seeds, a plant needs to receive pollen from a flower of the same species, which may be hundreds of feet (meters) away. Evolution has found many ways to affect this transport across distances. Many trees, particularly in temperate zones, rely on wind to carry their pollen to a receptive flower. Unfortunately,

this is a random process, and most of the pollen will land anywhere but on an appropriate flower. A more efficient method is the direct transport of pollen by animals that visit one flower after another. However, animals will not just altruistically offer their transport services; they need to be convinced that it is worth their time to visit flower after flower. Flowers therefore produce nectar—essentially sugar water—which is a valuable food resource. Additionally, flowers are shaped in such a way that nectar-seeking animals will inevitably touch the pollen-delivering stamens and the pollen-receiving pistil. The feeding animal will thus involuntarily transfer pollen between flowers, fertilizing them.

A wide range of animals pollinate plants, including small insects, birds, and mammals. Even though some flowers receive visits from a large number of animal pollinators, they still have a high proportion of pollen loss because opportunistic feeders visit a wide variety of flowers. More direct pollen transport is accomplished when access to a floral resource is limited to only a

The cup-sized balsa flowers (*Ochroma pyramidale*) open when no other trees are flowering and attract a wide range of pollinators, such as this greater spear-nosed bat (*Phyllostomus hastatus*). This bat will land briefly and quickly drink the nectar pooled at the bottom of the flower.

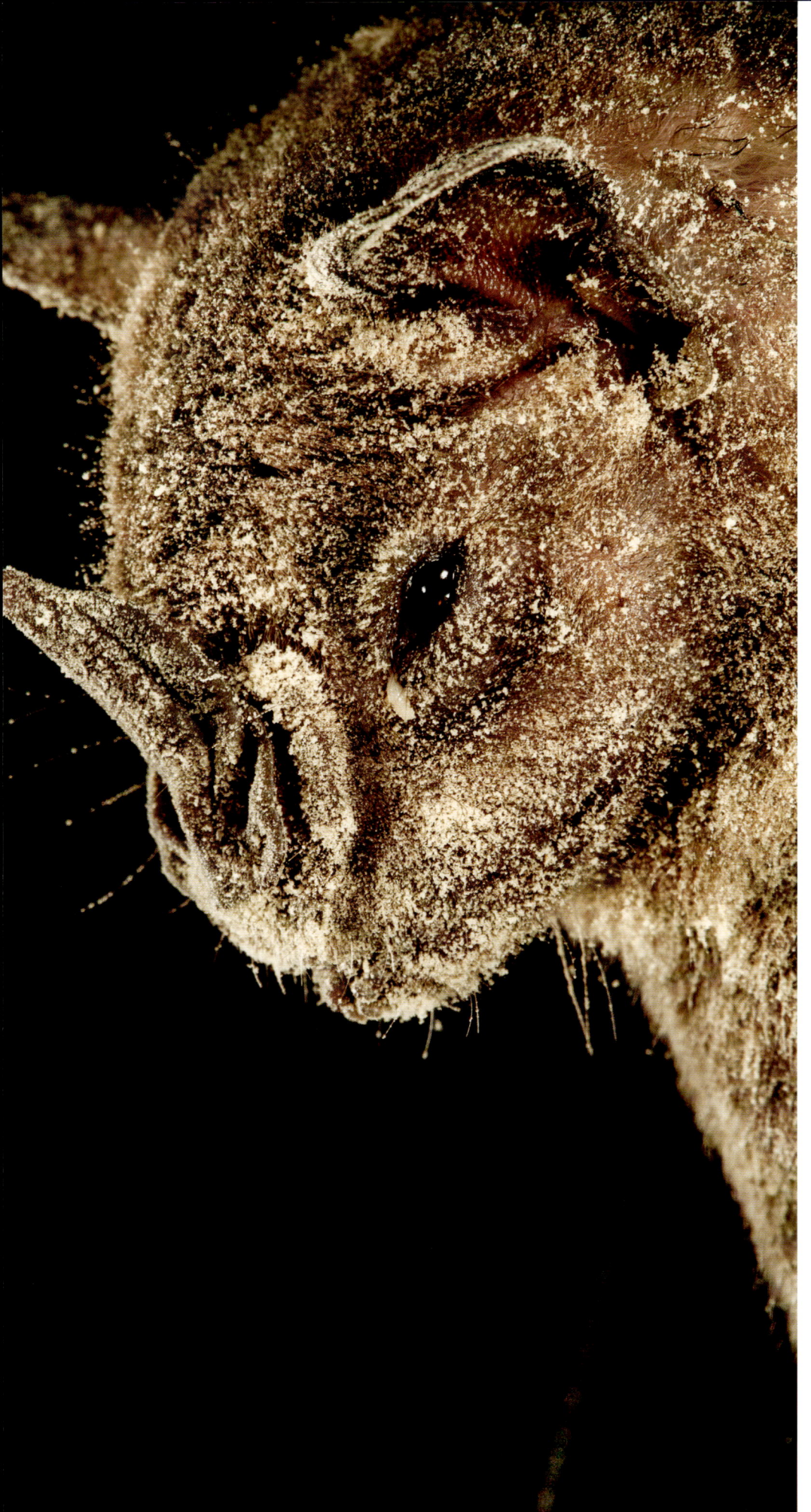

Some plants liberally dust pollinators with pollen while others strategically deposit smaller pollen packages on specific parts of the bat's body that are sure to come in contact with the next plant. Here, a Jamaican fruit-eating bat (*Artibeus jamaicensis*) is completely covered in balsa (*Ochroma pyramidale*) pollen. *A. jamaicensis* live predominantly on a diet of figs, but occasionally cannot resist big flowers that offer easy access to a lot of nectar, such as those of the balsa tree.

few species. Among the more specific flower-pollinator interactions are those that evolved between night-blooming flowers and nectar-feeding bats. Since bats are large in comparison to insects, bat-pollinated flowers are quite big—it takes a lot more nectar to satisfy a hungry bat than a small insect. Creating and provisioning a bat-pollinated flower is a large investment on the part of the plant. However, since bats fly longer distances than the average insect, they can transfer pollen very efficiently, even between plant individuals growing far apart. Evolutionarily, this investment pays off.

Since bats are active at night, bat-pollinated flowers are not characterized by the vivid colors that attract visually orienting pollinators during the day. Many bat-pollinated flowers are green, brownish, or pale and blend into the surrounding background vegetation. Remarkably, nectar-feeding bats are able to perceive ultraviolet light. Bat-pollinated flowers have been shown to reflect strongly in the ultraviolet spectrum. It remains to be seen to what extend bats use these wavelengths for locating their nectar resources.

To be detected more easily by bats, the majority of bat-pollinated flowers have peculiar smells that are often described as garlic-like, musky, or "like a cadaver." These smells, which are generally unpleasant to humans, are caused by sulfur compounds that only bats respond to, thus ensuring that the flowers are visited exclusively by bats.

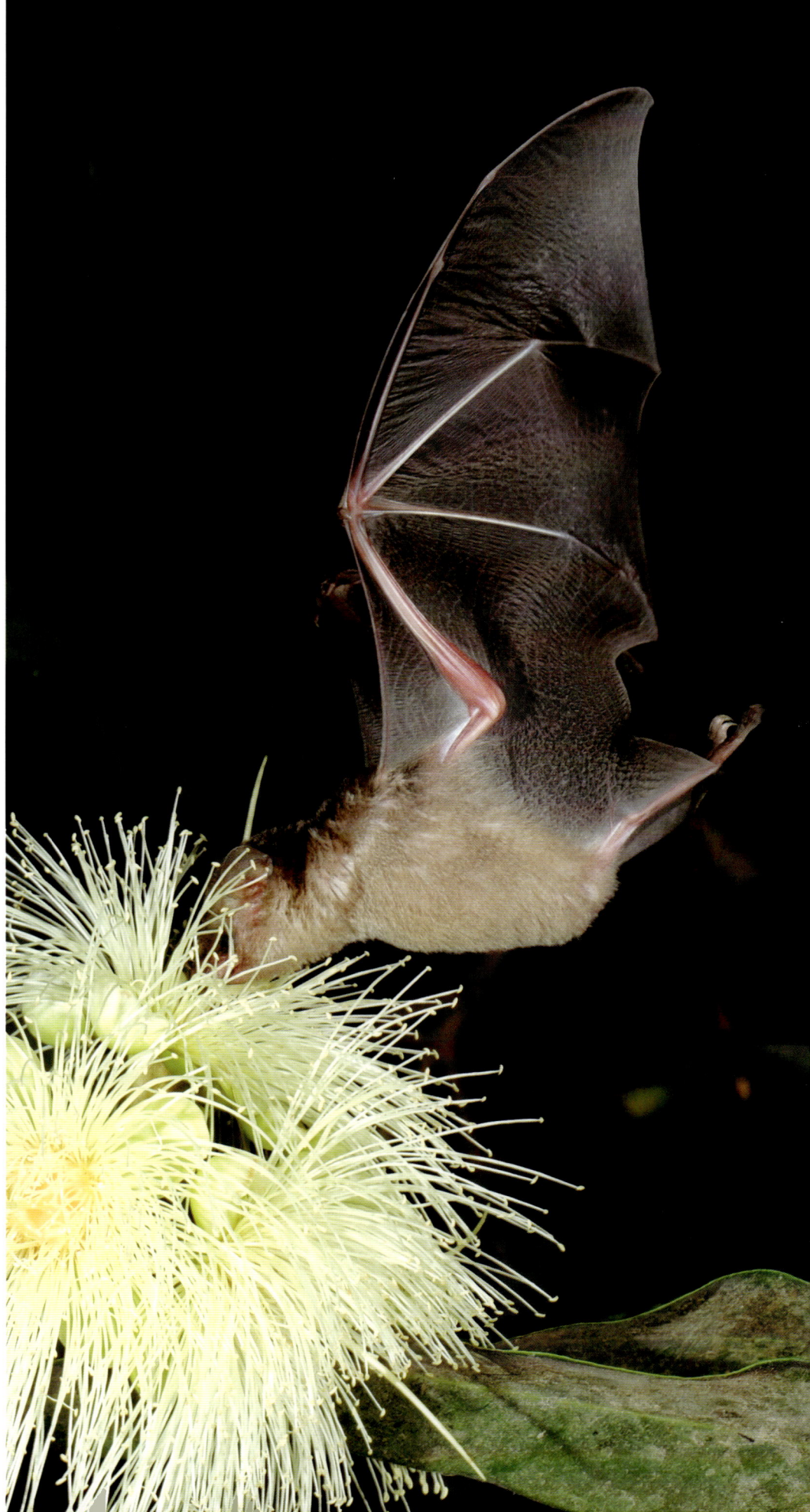

A Pallas's long-tongued bat (*Glossophaga soricina*) hovers briefly at a brush-like (*Syzygium* sp.) flower. Bat visits to flowers are extremely short and rarely last longer than a second. Photo © Marco Tschapka.

A few bat-pollinated flowers take advantage of bat echolocation by having structures that bounce an especially loud echo back to the bat. For example, the greenish-yellow flowers of the legume *Mucuna* have a special petal, the vexillum, that acts as an acoustic reflector and produces distinct echoes that stand out among the multitude of other echo-reflecting objects in the forest. Another method plants use to facilitate detection by echolocation is suspending clusters of flowers, or inflorescences, on long stalks below their branches, which results in a clearer echo for the bats, as the flowers are not obstructed by leaves. Some plants, such as the barrigón tree (*Pseudobombax septenatum*) avoid obstruction by leaves altogether by flowering during the dry season, a time of year when no leaves are present on this tree at all.

Plants flowering at the same time may share the same bat pollinator by coating different body parts of the bat with their pollen. For example, some plants cover the ventral side of the bat, while others deposit pollen on the bat's head, and still others on different areas—the *Mucuna* positions its pollen on the bat's uropatagium. While much of the pollen is transferred to other flowers, a portion is ingested by the bat when it grooms itself. These pollen snacks provide essential protein for nectar-feeding bats, supplementing their sugar-rich nectar diet.

Neotropical flower-visiting bats, such as Pallas's long-tongued bat (*Glossophaga soricina*), are in general rather small and have developed the capability of feeding from flowers while hovering in flight, similar to hummingbirds. Because of this, they do not need to land on the flower and are able to pollinate even small and delicate plants. To extract nectar from deep within a flower, flower-visiting bats have tongues that can extend out of their mouths to a length of over two-thirds their body size. They also have a characteristically elongated muzzle to accommodate such a long tongue. In most nectar-feeding bat species, the tip of the tongue is covered with hairlike papillae that efficiently soak up nectar like a mop before it is taken into the mouth. Species from another independently evolved group of nectar-feeding bats, including the orange nectar bat (*Lonchophylla robusta*), have deep grooves on both sides of the tongue with muscles that pump the nectar from the flower into the mouth.

The dietary habits of nectar-feeding bat species can be quite flexible and adaptable, depending on the resources within the bat's habitat. *G. soricina* prefers to forage on nectar from flowers. However, when no flowers are available, they may temporarily switch to feeding on soft fruits and some insects. There are also bats with a flexible diet that only occasionally visit flowers for nectar. One example of these opportunistic nectar-feeders is the greater spear-nosed bat (*Phyllostomus hastatus*), which feeds on insects and fruits as well as the nectar of particularly large flowers, such as those of the balsa tree (*Ochroma pyramidale*). *P. hastatus* is one of the largest bats in Latin America, with a wingspan of 24 inches (60 centimeters). These large bats do not hover, but instead drop deep inside the giant flowers of the balsa tree in search of nectar and emerge from the flowers completely covered with pollen.

Feeding on flowers involves mutualistic interactions that not only satisfy hungry bats but are also essential for plants. Plant reproduction does not stop at the production of seeds; the seeds then need to find a good place for germinating and developing into an adult tree. Bats also play an important role in this next step of plant reproduction, which we will see in chapter 9.

A Pallas's long-tongued bat (*Glossophaga soricina*) reaches for nectar hidden in a deer-eye bean (*Mucuna mutisiana*) flower. The raised, upper petal of the *Mucuna* flower acts like a reflector that provides a conspicuous echo target to the approaching bat. *Mucuna* flowers have an explosion mechanism triggered to release pollen at the first visit of a bat. The bat will cling briefly to the flower and the pollen will be deposited on its uropatagium.

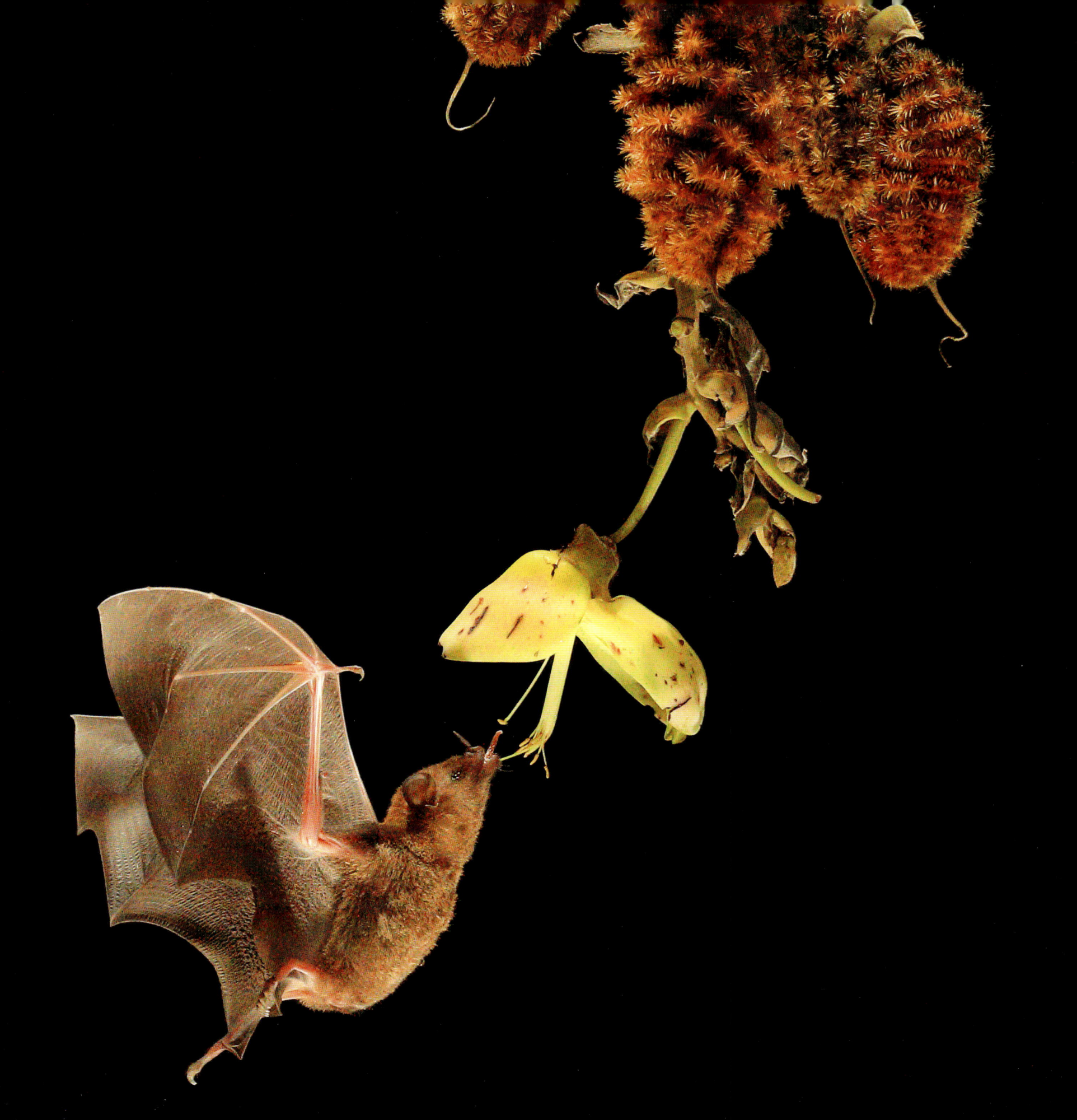

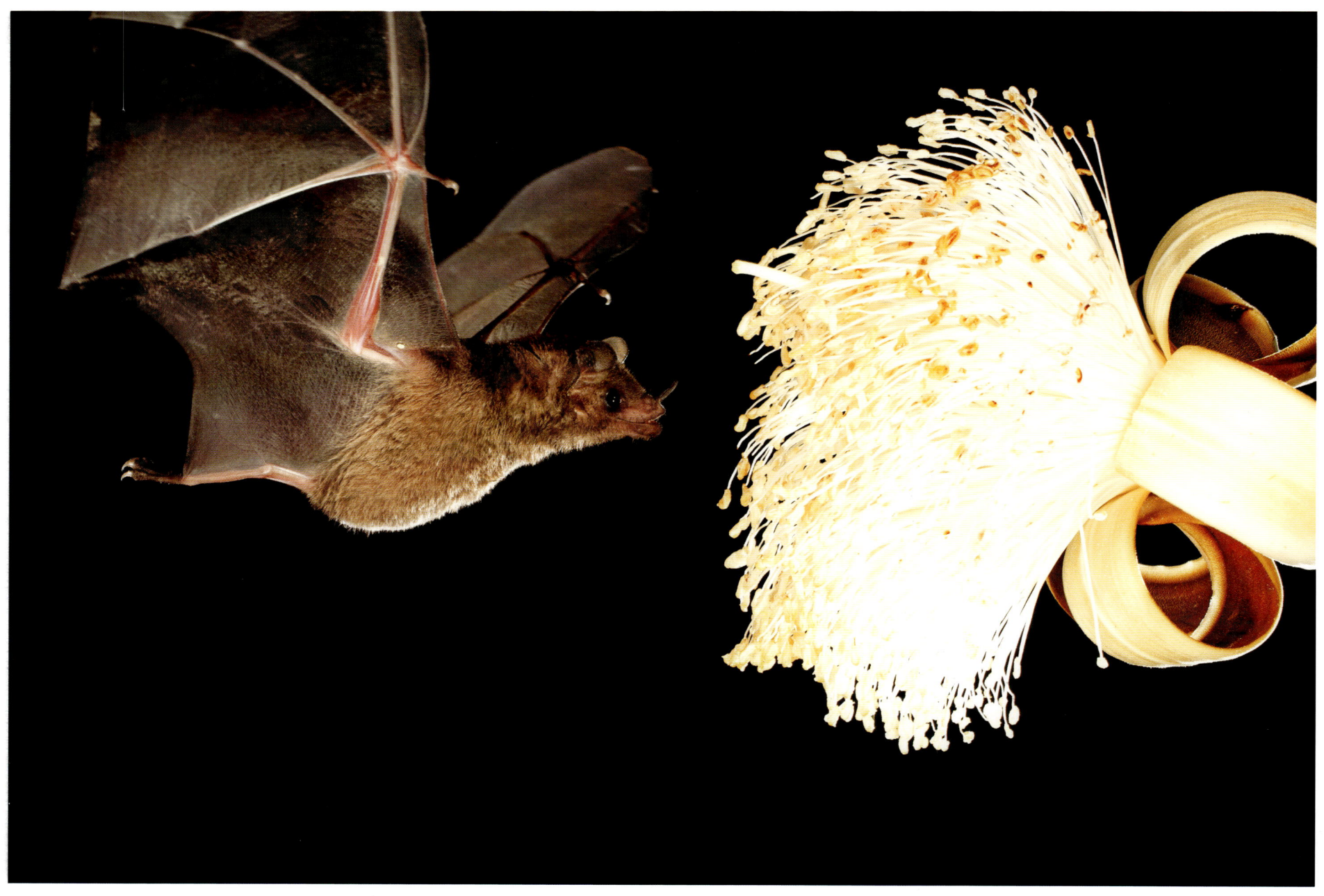

ABOVE AND OPPOSITE A Pallas's long-tongued bat (*Glossophaga soricina*) approaches a large flower (*Pseudobombax septenatum*) searching for a meal of nectar. To reach the nectar the bat must dive into the mass of stamens and becomes entirely covered with the pollen in the process.

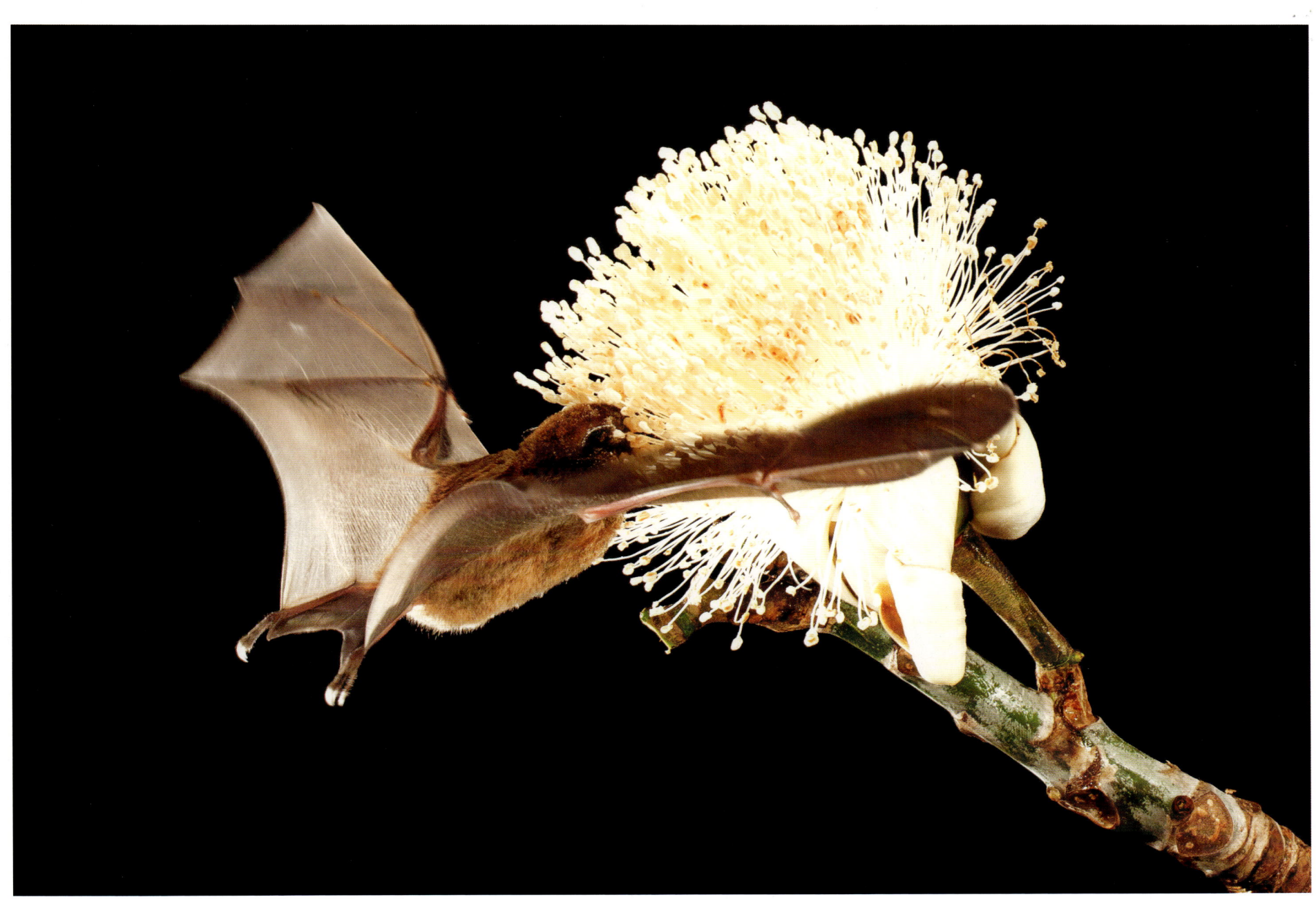

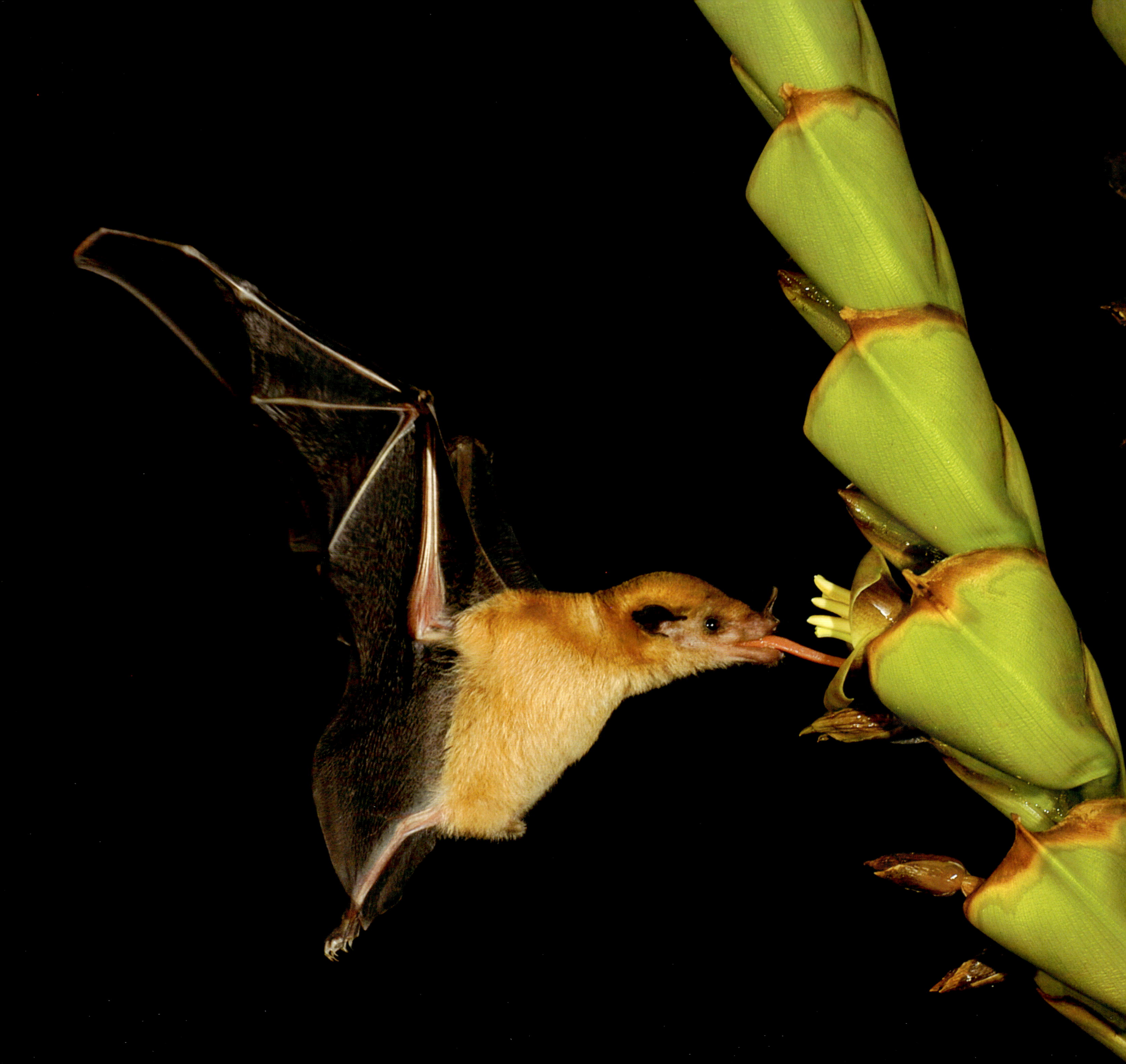

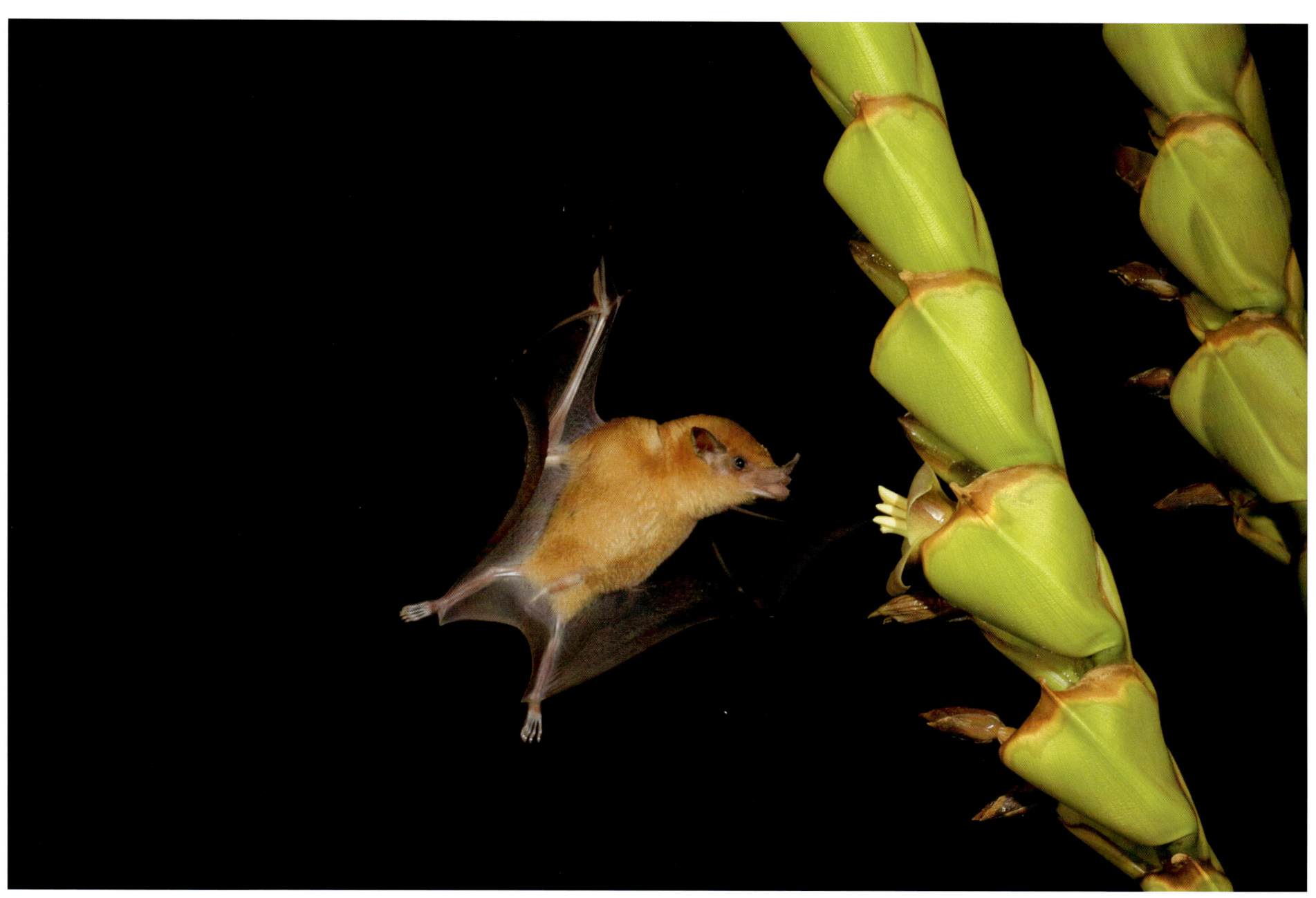

OPPOSITE AND ABOVE The orange nectar-feeding bat (*Lonchophylla robusta*) visits the flower of an epiphytic bromeliad (*Werauhia sanguinolenta*). To reach the nectar at the bottom of the flower, this bat has an extremely elongated tongue that can stretch to almost two thirds of its body length. Photos © Marco Tschapka.

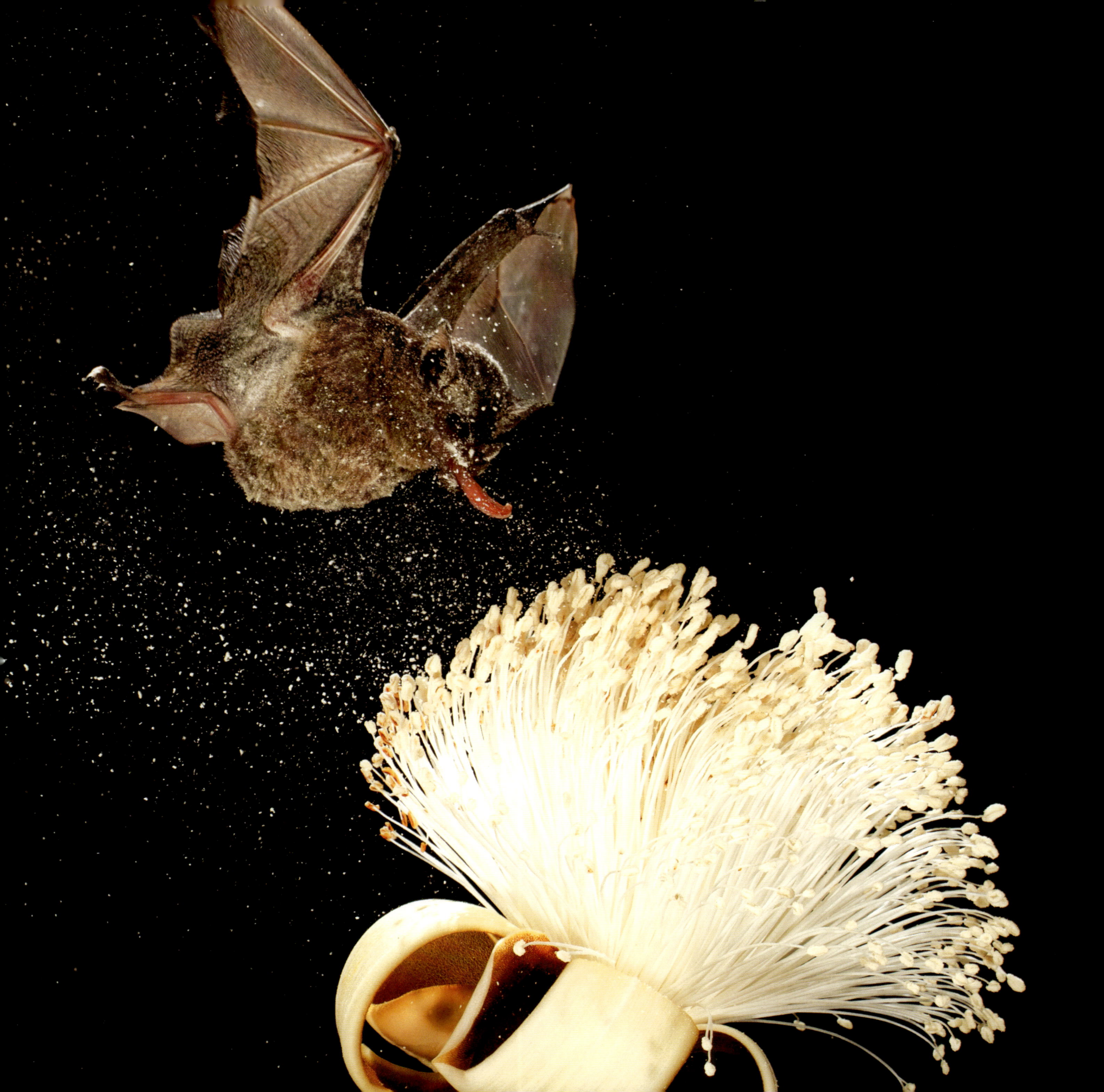

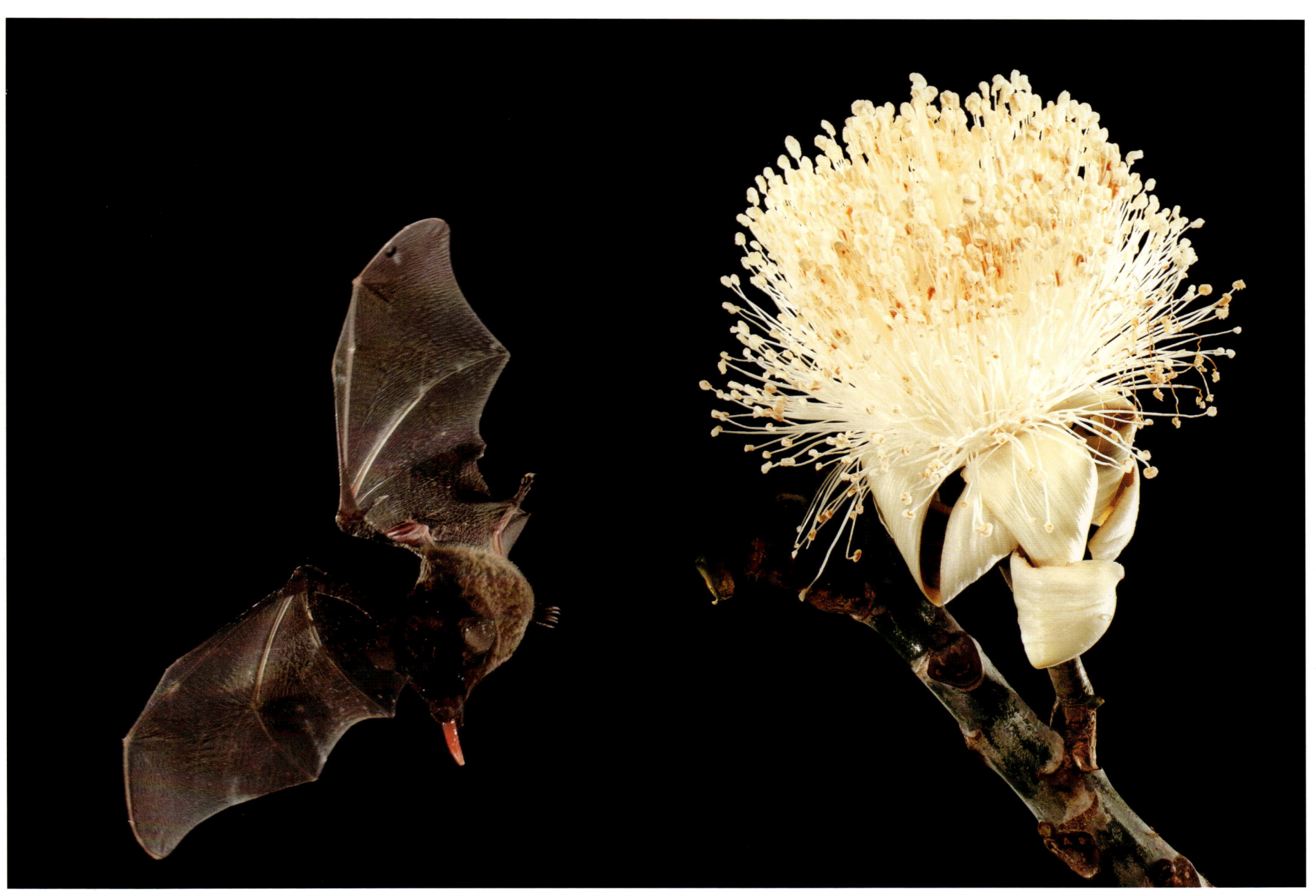

OPPOSITE AND ABOVE Flowers of the *Pseudobombax septenatum* tree provide enough pollen to shower visiting bats. After feeding at one tree, a Pallas's long-tongued bat (*Glossophaga soricina*) may move on to another, brushing the pollen on its fur into an awaiting flower. Any pollen left on its fur at the end of the night will be licked clean using its long tongue.

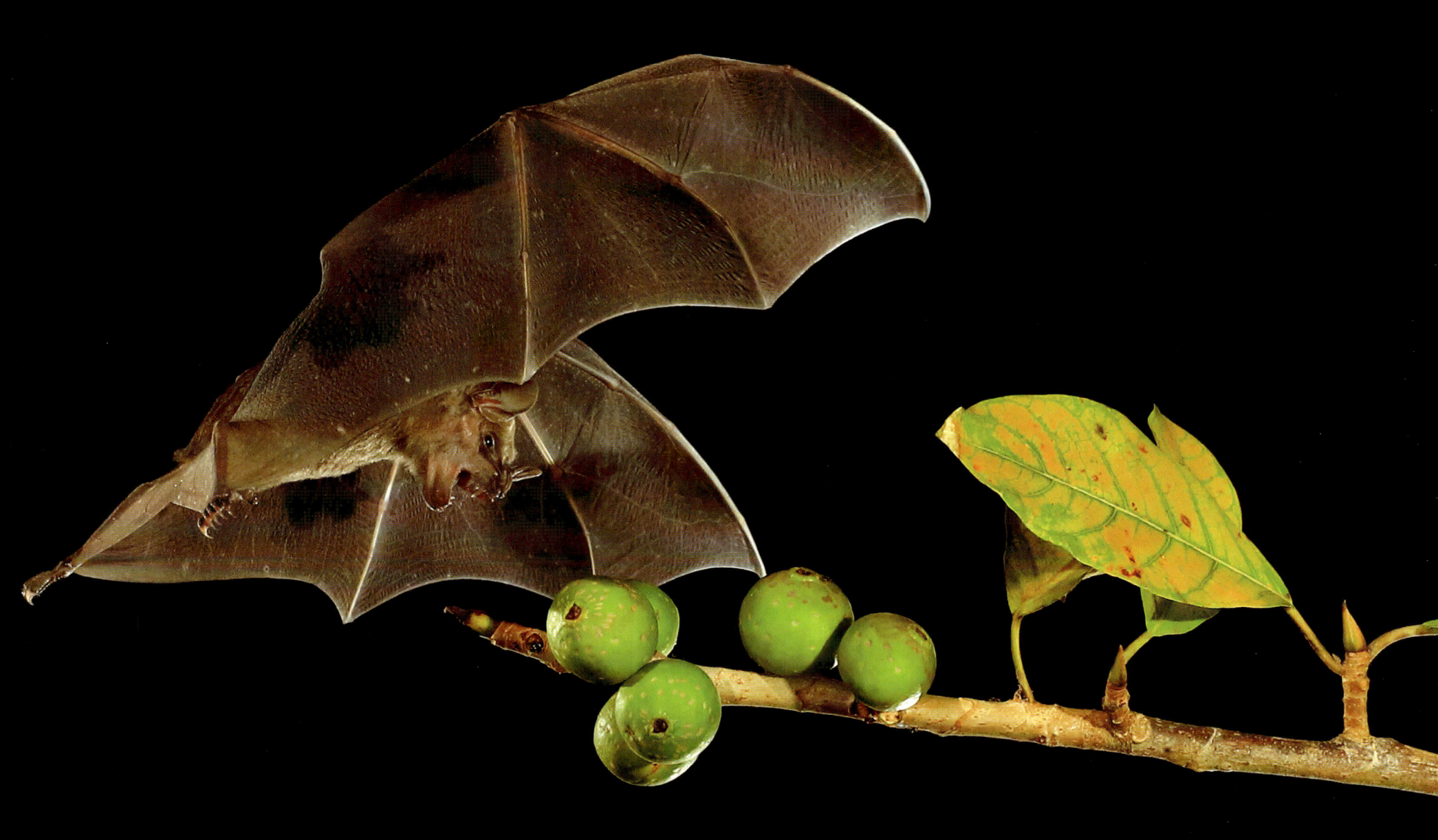

CHAPTER NINE

SEED-DISPERSING FRUIT LOVERS

FROM COOPERATION TO CHEATING

If all a plant's seeds simply fell to the ground near the mother plant, they would not have enough space and resources to grow into seedlings and ultimately into adults. To avoid such overcrowding, plants evolved fruits to envelop their seeds in a sweet fruit pulp that attracts hungry animals. Animals transport the fruits and seeds away from the mother plant. Since bats fly long distances during foraging, they are excellent seed dispersers that aid plant survival and promote genetic connectivity, even among dispersed plant populations.

As we saw in chapter 8, bat-dispersed fruits differ distinctly from those dispersed by other, day-active animals. While fruits dispersed by diurnal birds are often brightly colored, many bat-dispersed fruits are green and have strong smells. Examples of trees and shrubs with green bat-dispersed fruits are the genera *Cecropia*, *Solanum*, and *Piper*. Fig trees (*Ficus* spp.) present a particularly interesting case. Figs are a very important food resource for many frugivorous bats, as many fig species do not have a distinct fruiting season. Instead, each fig tree produces fruits on an individual schedule, asynchronous with the fruit production of other fig trees, such that there is always a fig supply for frugivores in a forest at any given time.

BCI harbors many species of fig trees that can be separated into two different groups based on their type of fruit: Some have red figs that have no smell and are distributed by birds, while others are green with a distinct smell. The green figs are regularly eaten by frugivorous bats, which find their fruits primarily by scent. Some of the most abundant bats on BCI, such as the Jamaican fruit-eating bat (*Artibeus jamaicensis*), depend on figs for the bulk of their food.

A Jamaican fruit-eating bat (*Artibeus jamaicensis*) approaches a branch of ripe figs.

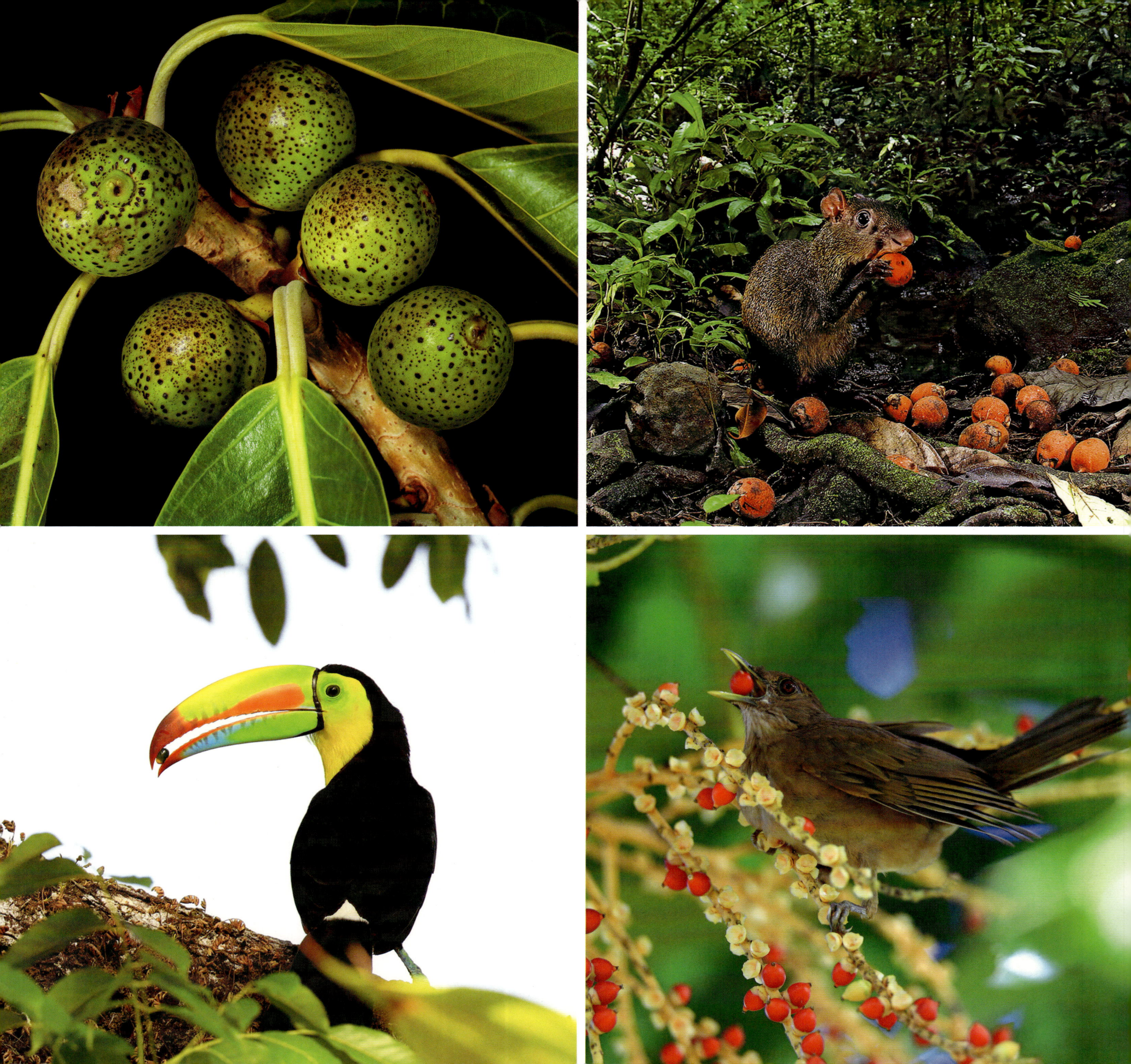

OPPOSITE CLOCKWISE FROM TOP LEFT Green figs (*Ficus insipida*) waiting to be dispersed by bats. Bats are not the only frugivores on BCI. Among others, agoutis (*Dasyprocta punctata*), clay-colored thrushes (*Turdus grayi*), and keel-billed toucans (*Ramphastos sulfuratus*) are also voracious fruit eaters. **BOTTOM LEFT** Red fig species are more commonly dispersed by birds and primates, as the bright red attracts visually oriented, diurnal species that use the conspicuous color to find the fruits. **BOTTOM RIGHT** A cross section of a ripe fig. Fruit-eating bats will chew these fruits into a pulp and squeeze out the juice to extract as much energy as possible. They then spit the dry fibrous pulp onto the forest floor to minimize the weight they carry in flight.

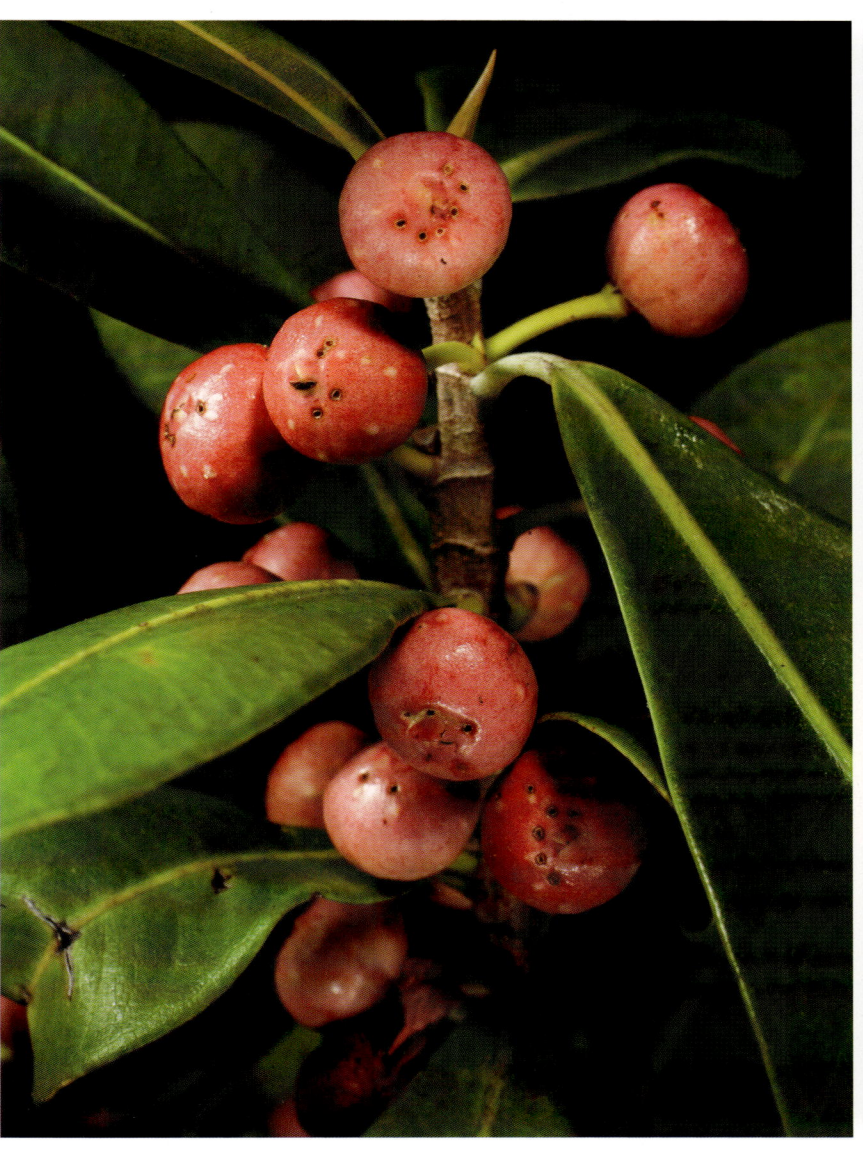

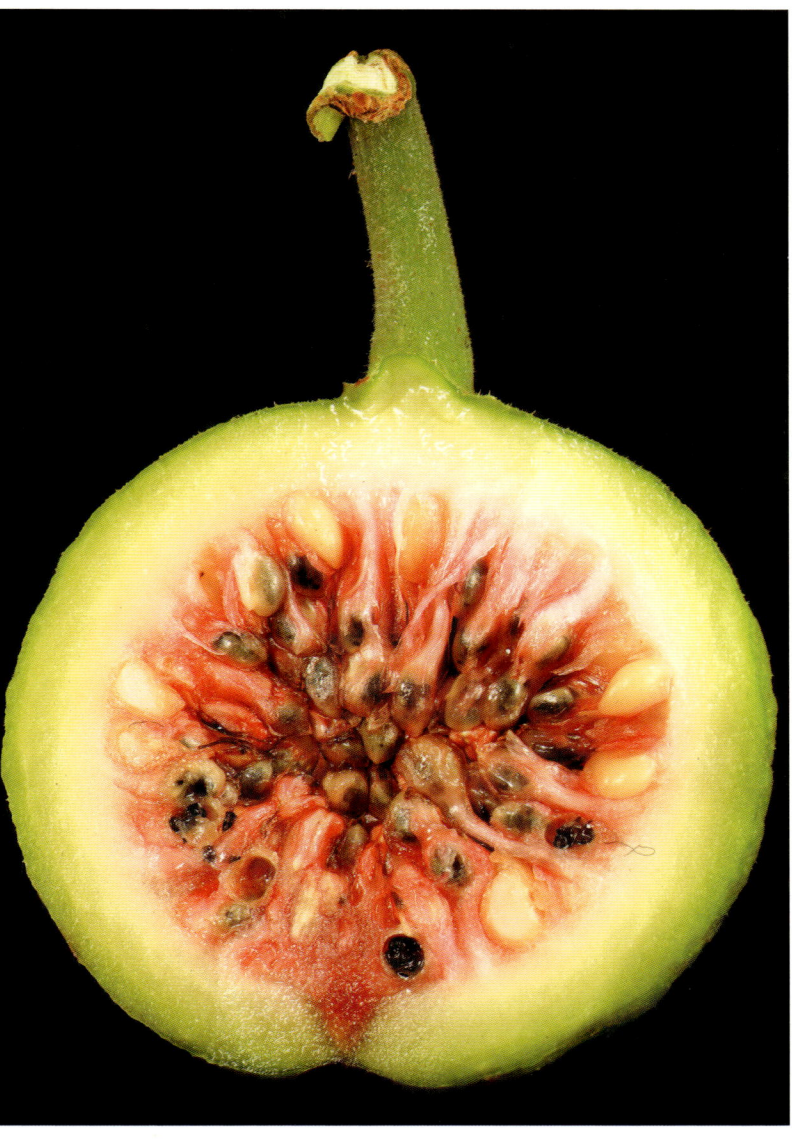

Another group of plants popular with bats is the pepper family Piperaceae. Many pepper species depend on bats for the dispersal of their seeds. They present elongated fruits that either dangle down or stand erect like Christmas tree candles on a branch. Short-tailed bats, such as Seba's short-tailed bat (*Carollia perspicillata*) and the chestnut short-tailed bat (*Carollia castanea*), are the most frequent feeders at these pepper plants. These bats rely on a combination of senses to find the fruits on the plants. Ripe fruits are initially detected by smell, and the final approach and harvest of the fruit is coordinated through echolocation.

Frugivorous bats are characterized by medium-sized ears and a short, broad rostrum (or snout). Their molars are wide and evolved to squeeze fruit juice out of the fruit pulp. A frugivorous bat bites a chunk off a fig and then chews it as long as there is a taste of sugar, like humans with chewing gum. The tasteless remains consist of indigestible fruit fibers, most of which the bat spits out to minimize the otherwise dead weight in flight. While chewing the fruit chunks, the bat also inevitably ingests the small seeds hidden in the pulp. The seeds pass through the digestive tract and exit the body intact with a small amount of fertilizer when the bats defecate.

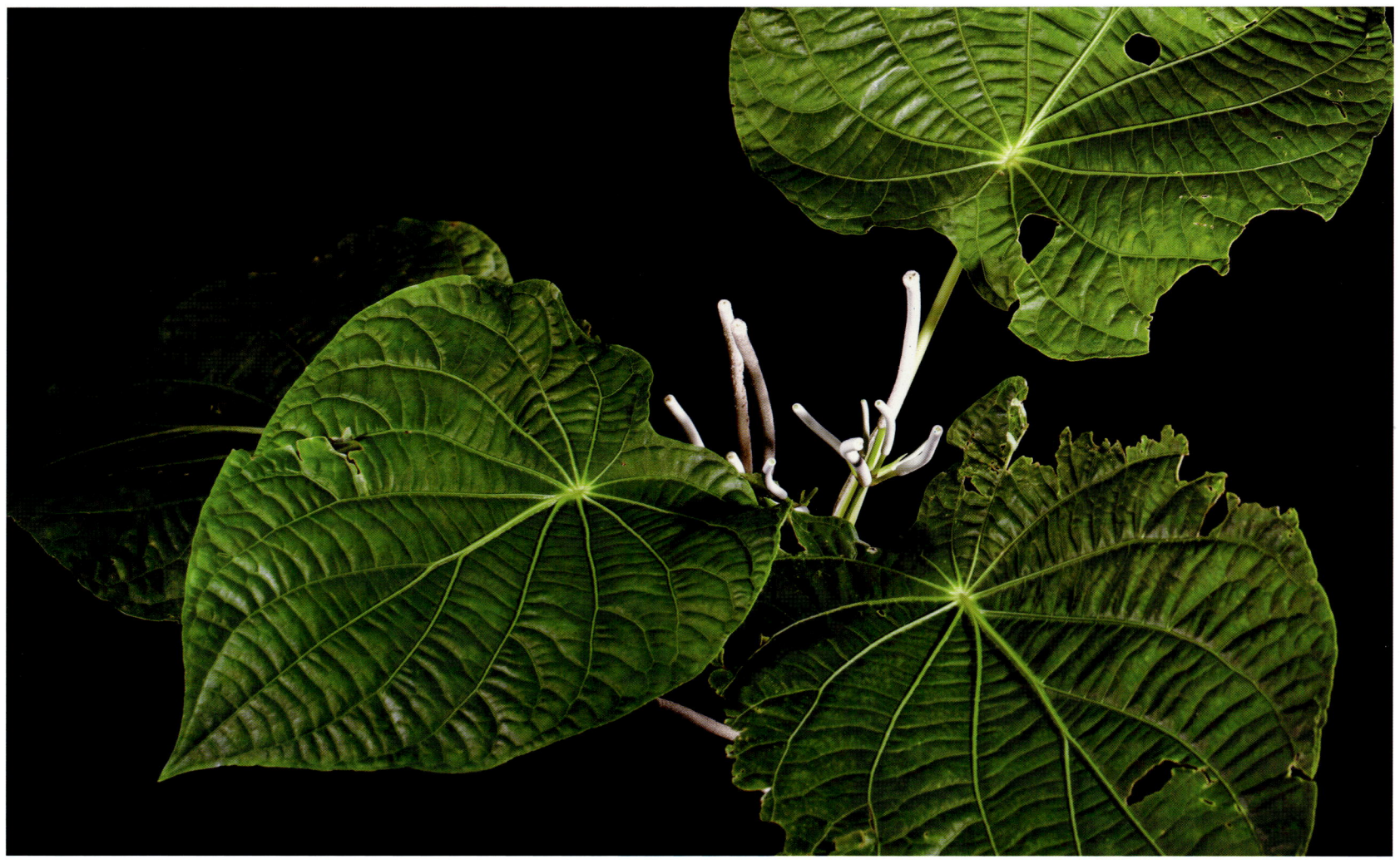

Fruit from Neotropical pepper plants (*Piper* spp.) comprise the bulk of the diet for short-tailed bats (*Carollia* spp.). Some short-tailed bat species are true gourmets and only feed on specific *Piper* species, while others also feed on many other fruits in the forest.

Other plants produce fruits with seeds too large for bats to swallow and transport within their intestines. In these cases, bats carry the fruit away from the fruit plant in their mouths, and then hang in another location, eating all the edible parts of the fruit until only the seed is left, which is then dropped to the forest floor. Both of these methods of seed transport effectively disperse the seeds a good distance from the mother tree.

Fruit bats on BCI come in all sizes, ranging from the tiny 0.2-ounce (6-gram) MacConnell's bat (*Mesophylla macconnelli*) to the 2.3-ounce (65-gram) great fruit-eating bat (*Artibeus lituratus*).

Their dietary interests are distributed over the size range of the available fruits. The smallest fruits are eaten, and their seeds distributed, by the small bat species, since small fruits are the only ones they are able to carry. In contrast, larger bat species are less interested in the smaller fruits, as it would be inefficient to find and eat enough of them to satisfy their hunger. They focus on larger fruits. Bats use scent and echolocation to identify fig species and prefer figs that correspond to their body size. This is one of many ways that the bats of BCI partition resources, allowing the coexistence of so many species.

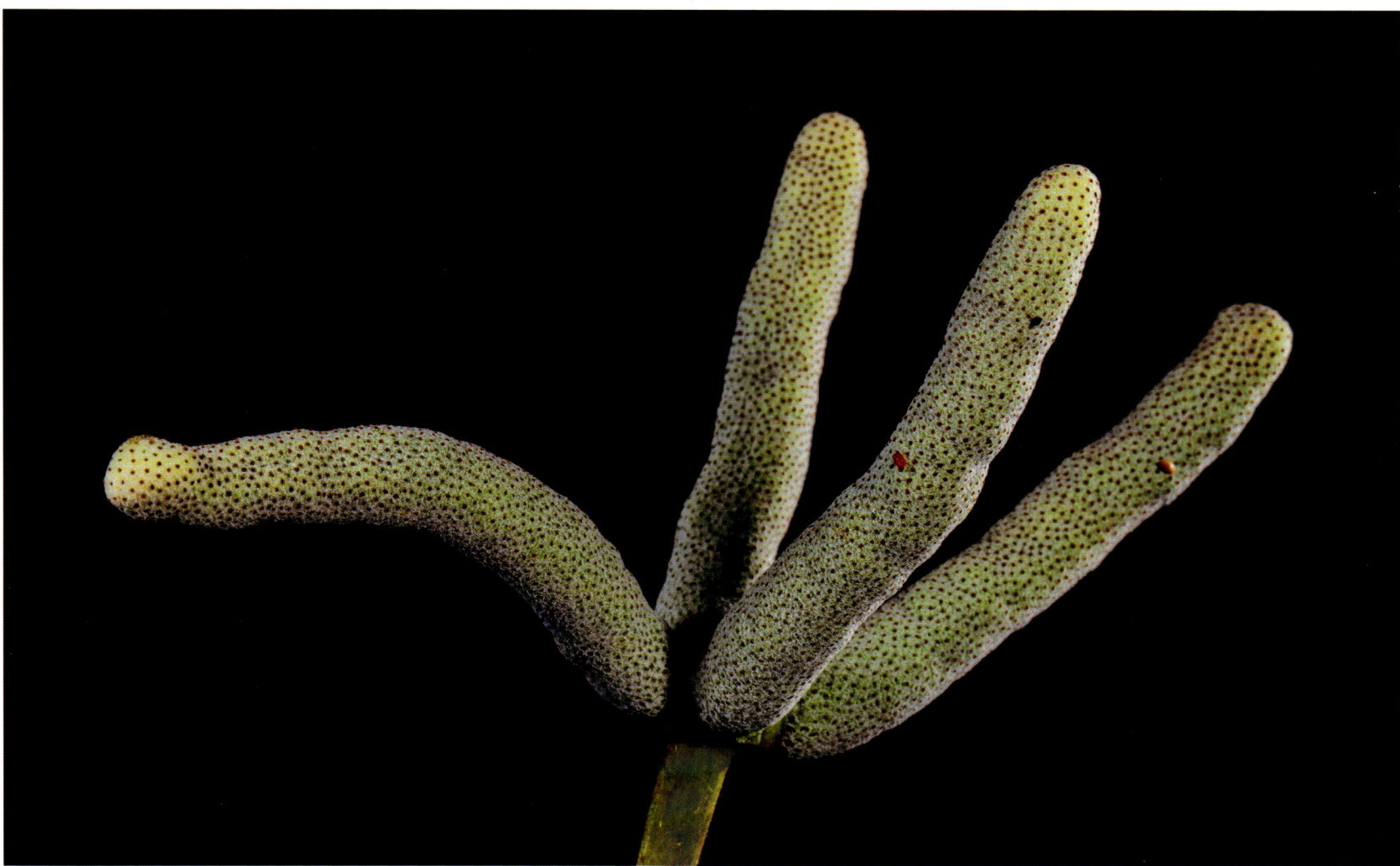

Fruits of the pioneer tree genus *Cecropia* are well-loved in the forest, including by bats that disperse their seeds into new areas. These fast-growing trees are some of the first to appear in gaps and at the forest edge and are key to building diverse forests.

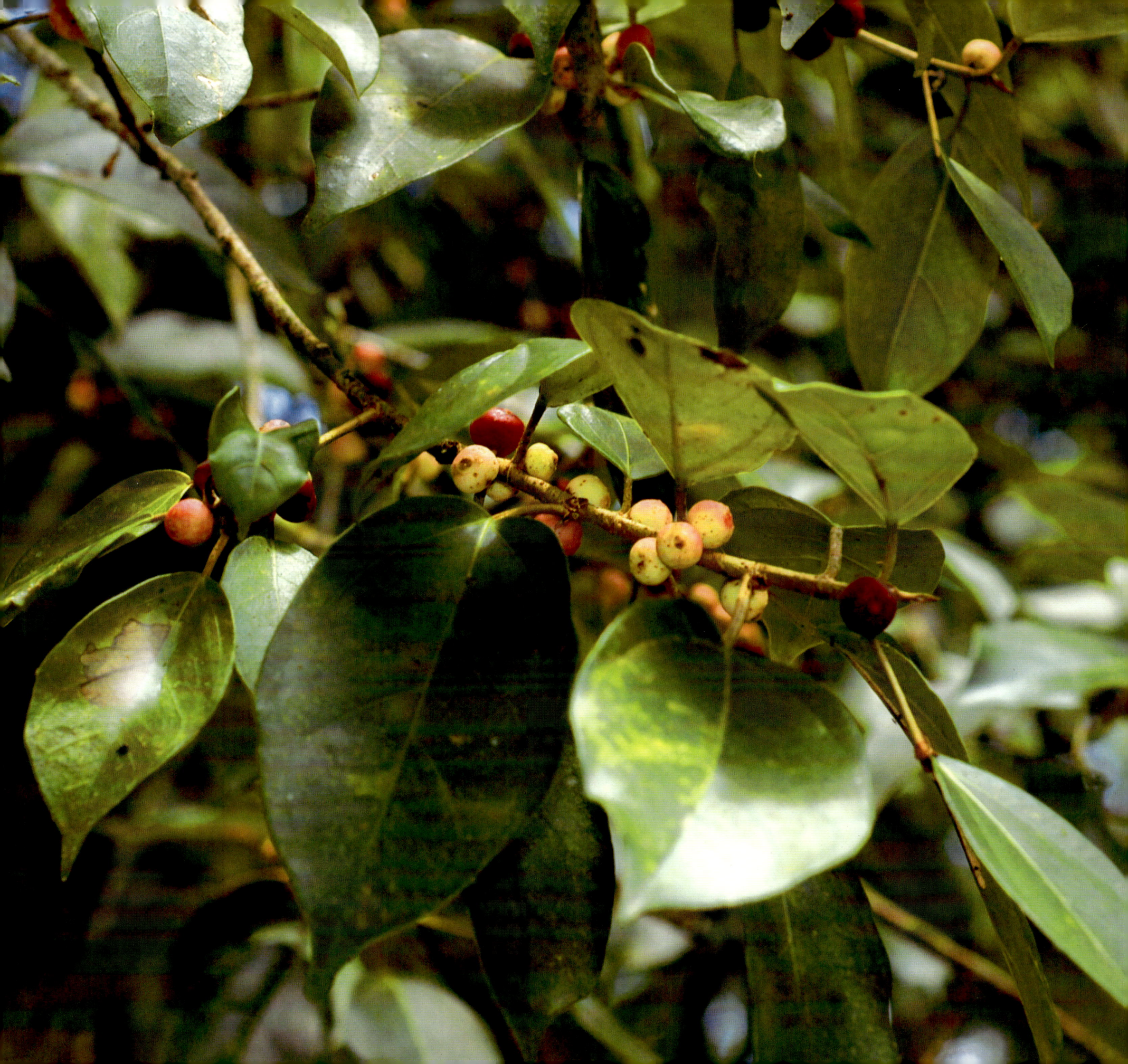

In general, the mutualistic exchange of seed dispersal for food works extremely well across many bat and plant species. There are bat species that cheat the system, however, to the detriment of the plants. The medium-sized hairy big-eyed bat (*Chiroderma villosum*) is an example. This bat feeds on larger figs, behaving initially just like other frugivorous bats. It chews small mouthfuls of a fig, extracts the juice, and spits out the remaining fiber packages. However, instead of swallowing the seeds whole and carrying them to a place where, after it defecates, the intact seeds can germinate, *C. villosum* chews the seeds and grinds them into small fragments with its powerful jaws to access the nutrients the plant had intended for its seedlings. While this is an evolutionary dead-end for the plant, luckily there are many more seed dispersers available to distribute the fig's seeds. A few cheaters can be tolerated within the seed-disperser system without significant losses for the plants. After all, only a few surviving seedlings are necessary to guarantee the genetic persistence of a tree.

Plant-visiting bats are essential for the reproduction of many rainforest plants, while in turn, bats depend on plants for food. The relationships between bats and their food plants form complex networks that define tropical rainforests. In the context of conservation this also means that the fates of the bats, and of their plant partners, are inseparably intertwined. When we lose bats, we will also lose the plants to which they are linked.

Red-fruited figs often rely on birds for seed dispersal, but this fig (*Ficus colubrinae*) also attracts small tent-making bats to feed on its blueberry-sized fruits and move its seeds through the forest. Photo © Marco Tschapka.

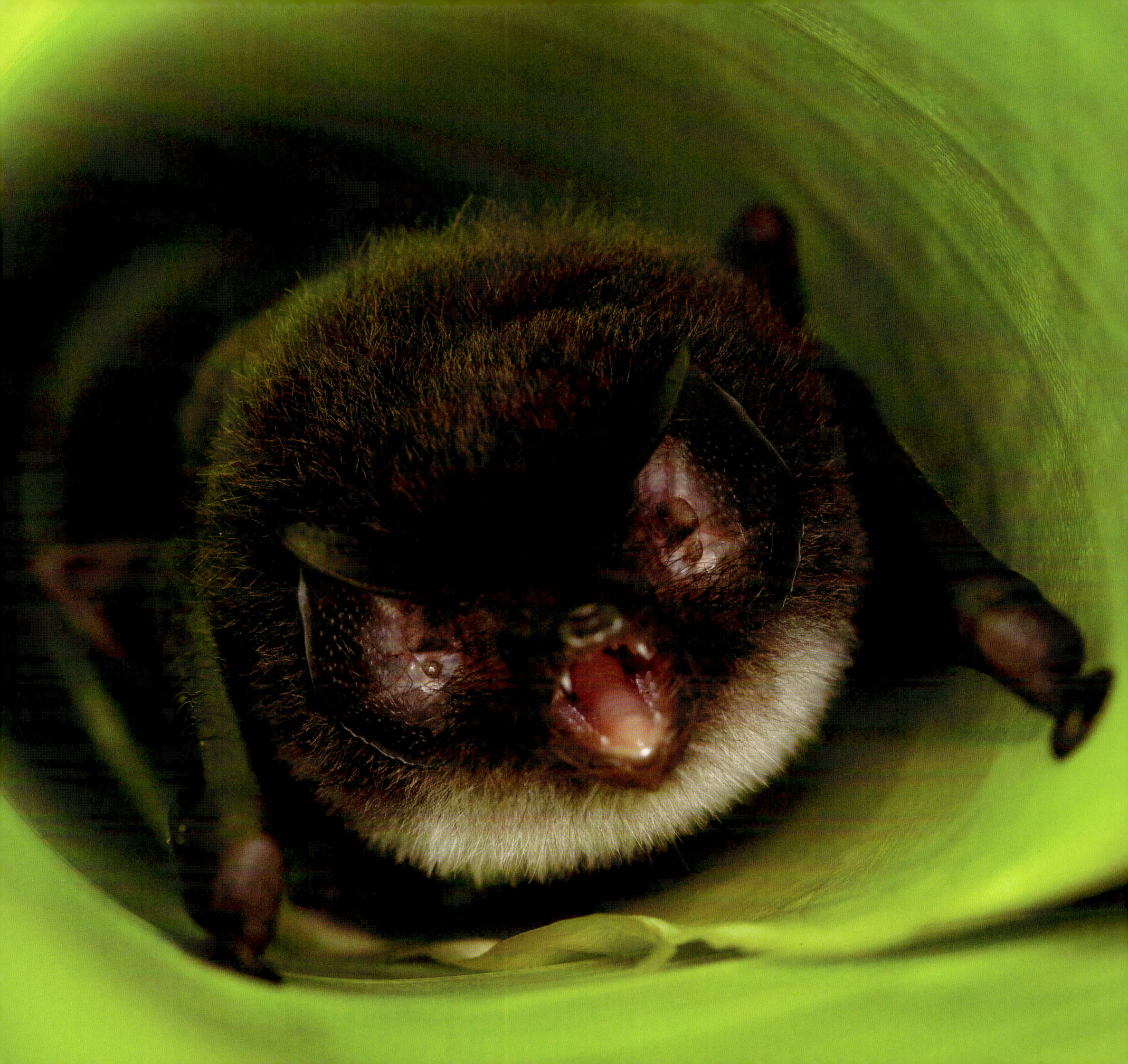

CHAPTER TEN

BAT ROOSTS

FROM OPPORTUNISM TO CONSTRUCTION

No matter how well adapted bats are at finding diverse food resources, they face the additional challenge of finding a safe place to hide and sleep during the day. Animals spend 50 to 90 percent of their time in their shelters. In addition to sleeping, shelters are also where many animals give birth and raise their young. Bats often have specific demands when it comes to their roosts. For example, the shelter has to be safe from predators, the microclimate has to be just right, and the entrance has to be accessible to flight. Bats are as diverse in the types of roosts they use as they are in their morphology, behavior, and diet. Roost availability is often limited, which is one reason certain bat species occur in some places but not in others. On BCI, for example, there are no caves or rock faces with crevices, so species only able to roost in these structures are absent from the island. One thing the BCI has many of is trees. Trees provide numerous possibilities for roosting bats. The greater sac-winged bat (*Saccopteryx bilineata*), for example, roosts among the buttresses at the bases of large fig trees and on tree trunks. As its name suggests, *S. bilineata* has two white lines that extend along the dark fur of its back, which allow it to blend in with the structures on the tree bark and conceal it from predators in the dappled light passing through the forest canopy. This disguise allows it to roost in relatively open places where it has enough room for its elaborate courtship displays. Another member of the sac-winged bat family, the proboscis bat (*Rhynchonycteris naso*), uses similar camouflage patterns. These bats roost one above the other in small groups on the underside of trees leaning over the water. With the coloration of their fur, they resemble pieces of bark, making them difficult to spot even for an experienced bat researcher.

A tiny Spix's disk-winged bat (*Thyroptera tricolor)* roosts in a furled leaf and calls to its group mates to join it. The openings of these roosts are only the right size for a single day, and disk-winged bats use call-and-response vocalizations to recruit group members to the next day's roost.

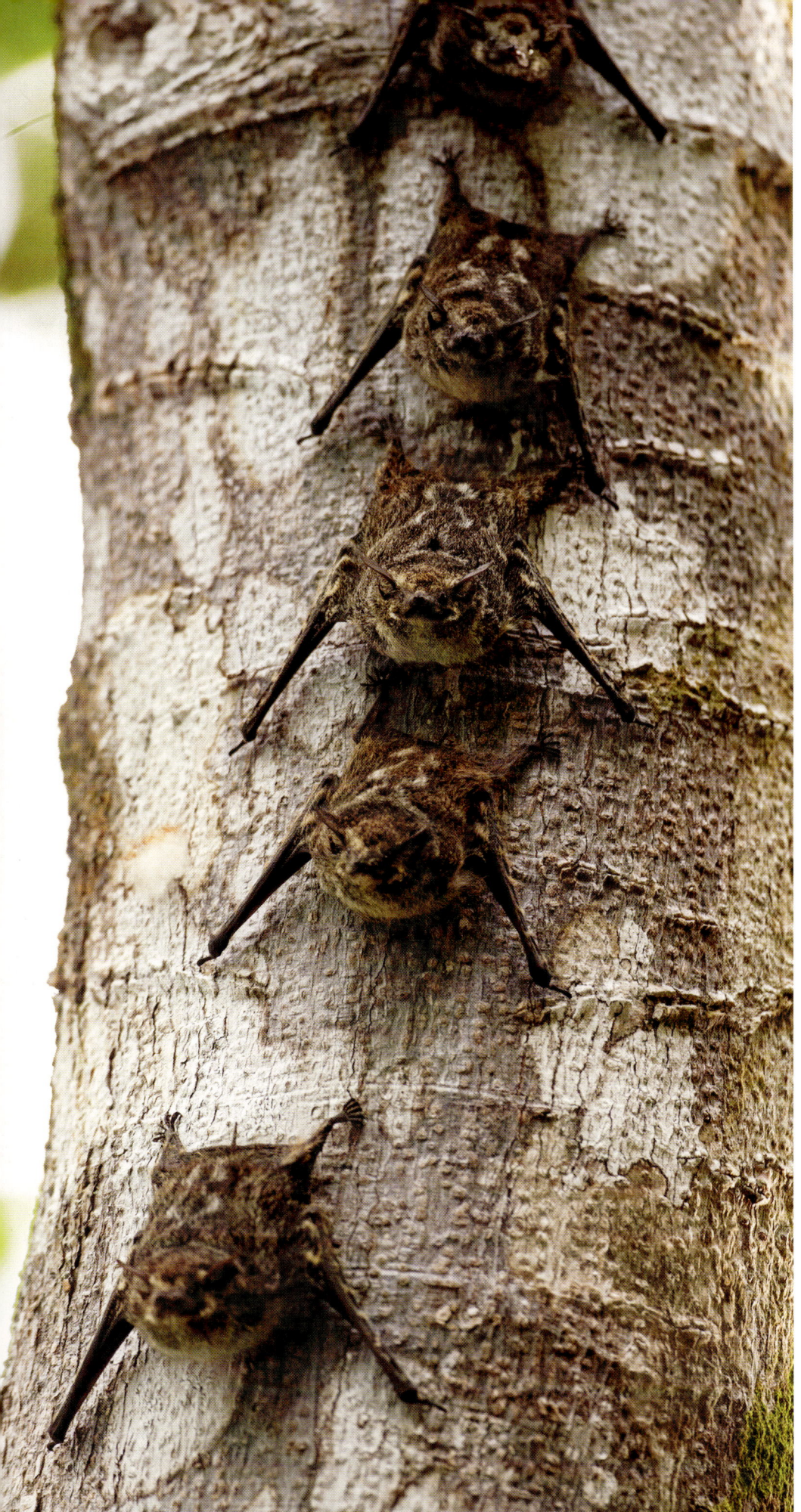

Some bat species roost in the open and rely on camouflage to avoid detection by predators. These proboscis bats (*Rhynchonycteris naso*) are roosting on a palm trunk next to a lagoon. Their mottled fur and zigzag stripes create nearly perfect camouflage from visual predators.

Most bat species prefer more sheltered roosts. Probably the most commonly used types of roosts on BCI are tree cavities. These roosts are relatively abundant, easy for bats to find, and provide a fairly good buffer from fluctuations in humidity and temperature. Tropical trees commonly have rotten cores, and the many old trees on BCI are no exception. It has even been hypothesized that it is adaptive for the tree to not invest in maintaining its core, but rather to let this be a site for animal roosts, such that feces build up and microbes metabolize them, supplying the tree with much needed nutrients in the tropics' nutrient-poor soils. Whether adaptive for the tree or not, these tree hollows provide bats with a variety of excellent roosting possibilities, with several bat species showing similar roost preferences. The stripe-headed round-eared bat (*Tonatia saurophila*), the fringe-lipped bat (*Trachops cirrhosus*), and the hairy big-eared bat (*Micronycteris hirsuta*) are examples of species that roost in tree cavities, preferably those with large openings at the base of the tree that are easy to fly into and out of. Some species have roosted in the same tree holes for so long and in such large numbers that it seems that the fumes from their accumulated droppings have turned the bats an orange color. On BCI, for example, the greater bulldog bat (*Noctilio leporinus*), a species that often is a dull gray-brown color, has turned bright orange.

Bats roost in trees during all stages of the tree's life and death, including in decaying stumps. A greater sac-winged bat (*Saccopteryx bilineata*) flies back to the roost after an early night of foraging for insects.

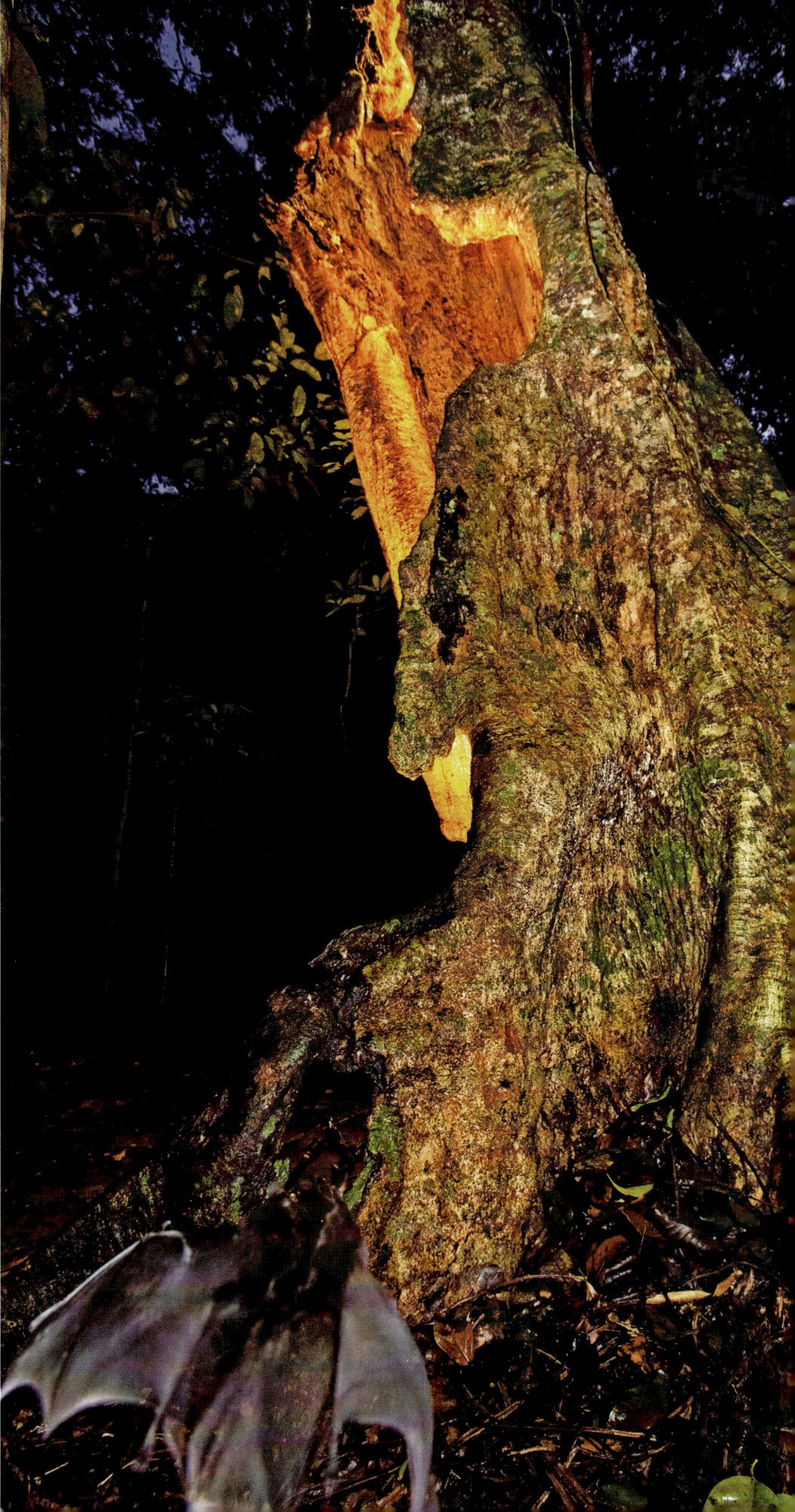

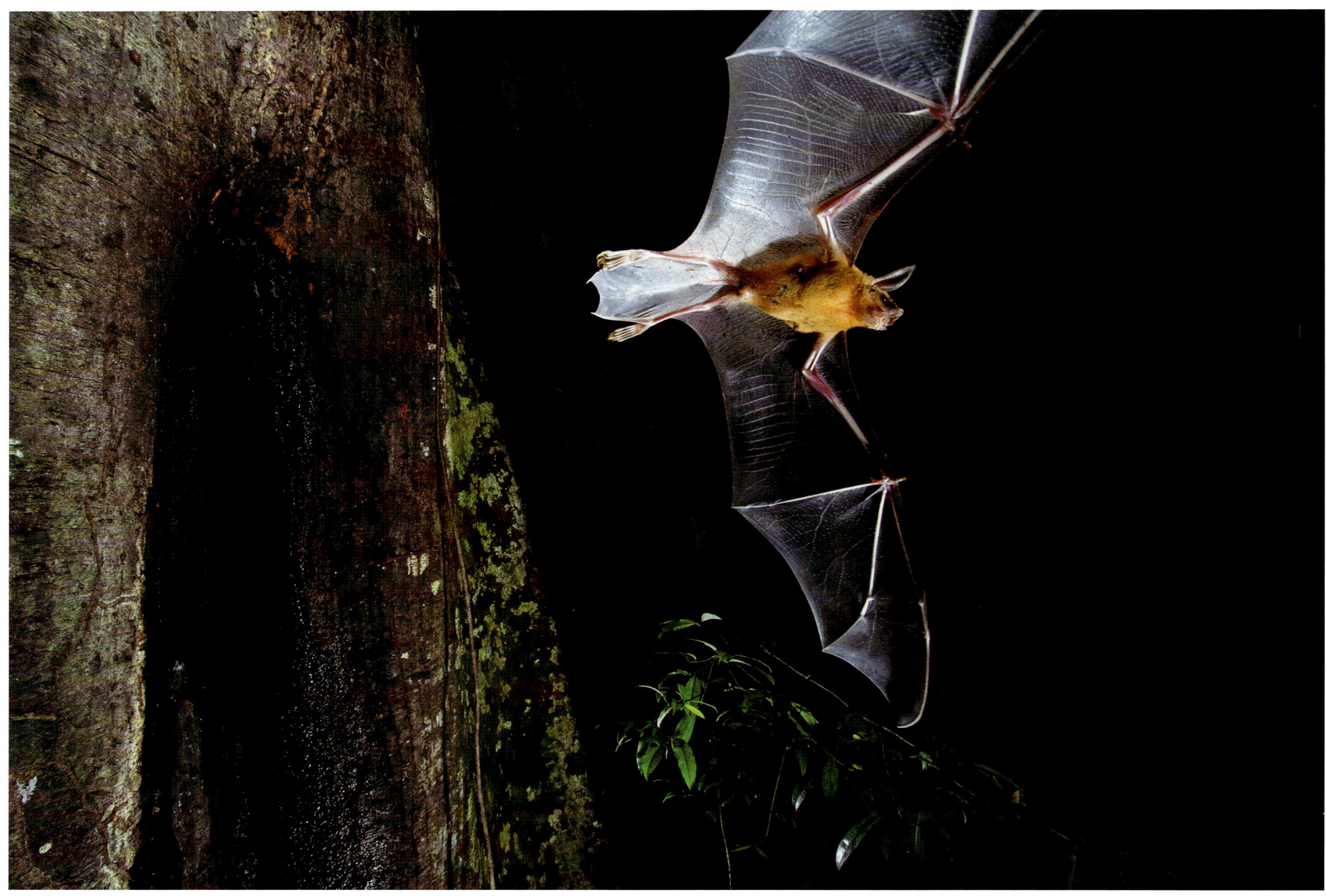

Old trees are crucial bat habitats. Many bat species rely on cavities formed when older trees begin to decay, creating openings at their base and a hollow roosting structure in the trunk. A greater bulldog bat (*Noctilio leporinus*) emerges from its day roost in a tree cavity.

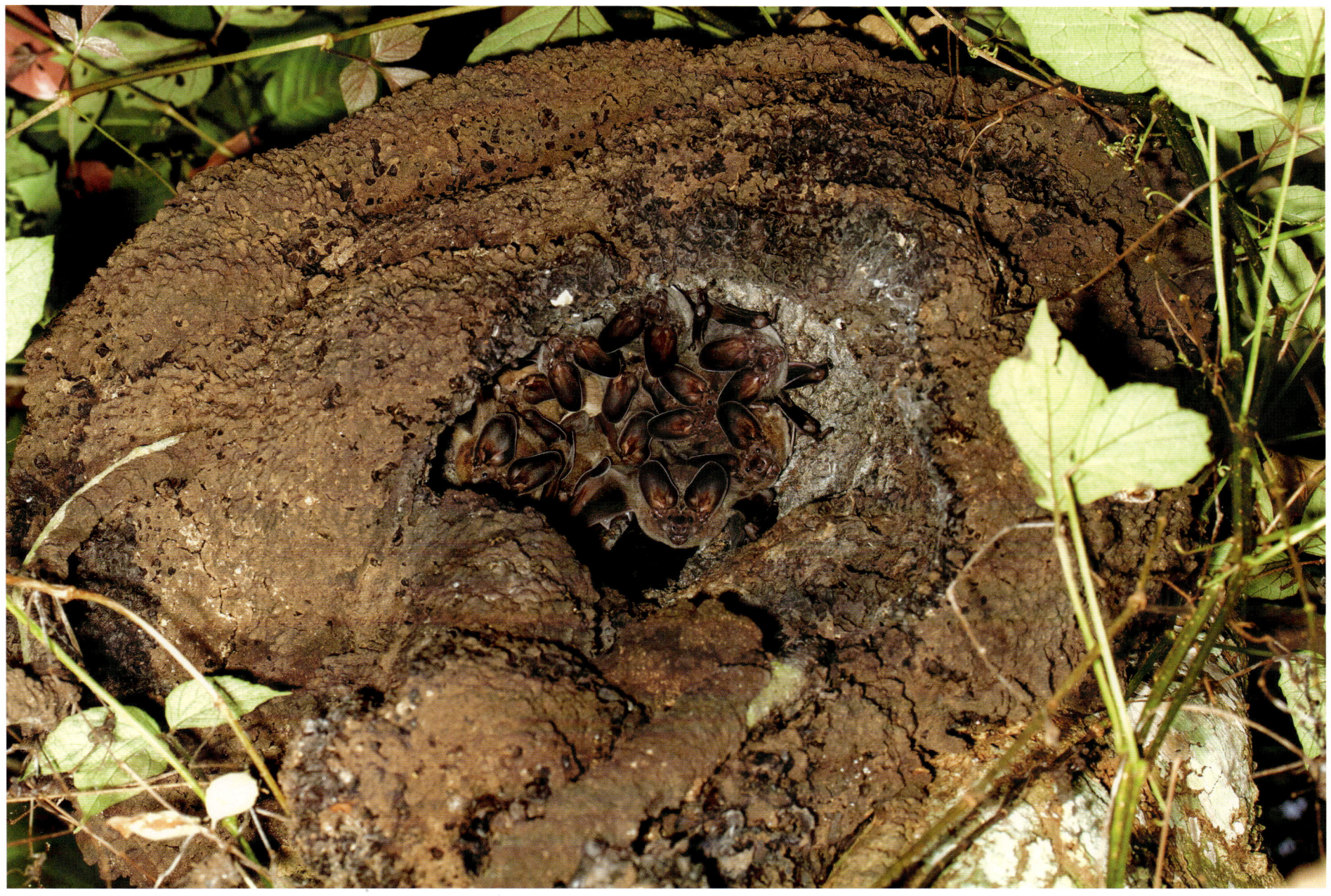

A group of white-throated round-eared bats (*Lophostoma silvicolum*) look down from their roost in an arboreal termite nest. A single male uses his teeth to hollow out a cavity in the bottom of the termite nest, fanning his tail membrane against the cavity wall between bites and likely spreading a chemical deterrent to keep termites at bay. Females then join to mate with him and raise their pups. Only active termite nests are used by the bats; the presence of the termites keeps the roost warm and protects the inhabitants from predators and parasites.

The largest bat species on BCI, the spectral bat (*Vampyrum spectrum*), with a wingspan approaching 3 feet (1 meter), also lives in tree holes. This species is not only a top predator, feeding on small vertebrates including bats, but is also thought to be monogamous, with just one male, one female, and sometimes some of their offspring occupying a single tree hole. When bat researchers caught a young female *V. spectrum* bat on BCI, they fed her a very large lizard before they let her go with a transmitter glued to her back. She led them to a tree hole that she did not leave for the next three days, digesting her meal like a lion after a big kill. *V. spectrum* is one of many bat species that remain elusive, and one that scientists would love to learn more about.

Tree cavities can provide much information for bat researchers. Insect-eating bats roosting in these cavities often bring prey home to eat in peace, and they frequently drop indigestible parts of the insects, mainly those composed of hard chitin such as wings and legs. In one study on BCI these insect remains were collected for a year below a roost of the hairy big-eared bat (*Micronycteris hirsuta*). Surprisingly, many ovipositors (tubelike structures for laying eggs) from katydids were discovered, indicating that the bats not only eat singing male katydids, which they can detect by their noisy songs, but also the silent females. This discovery changed our understanding of bat foraging behavior and the interactions between bats and katydids.

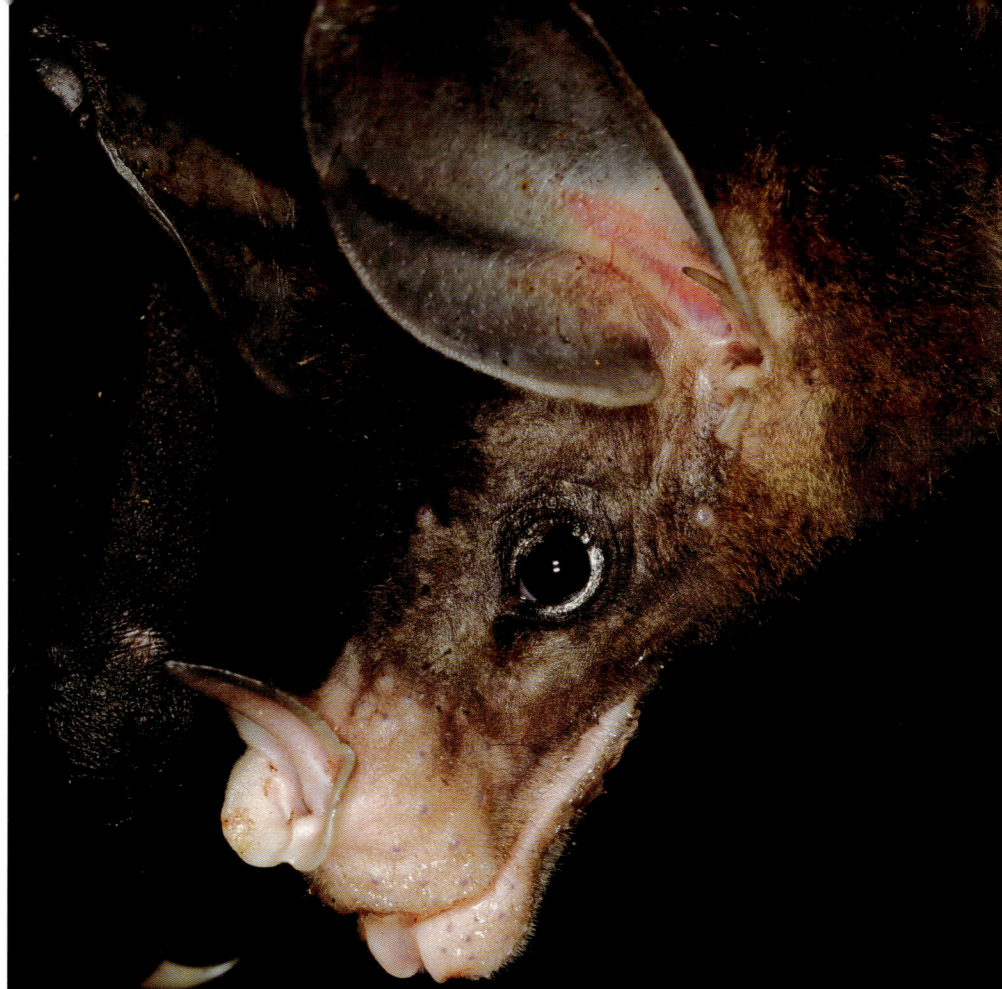

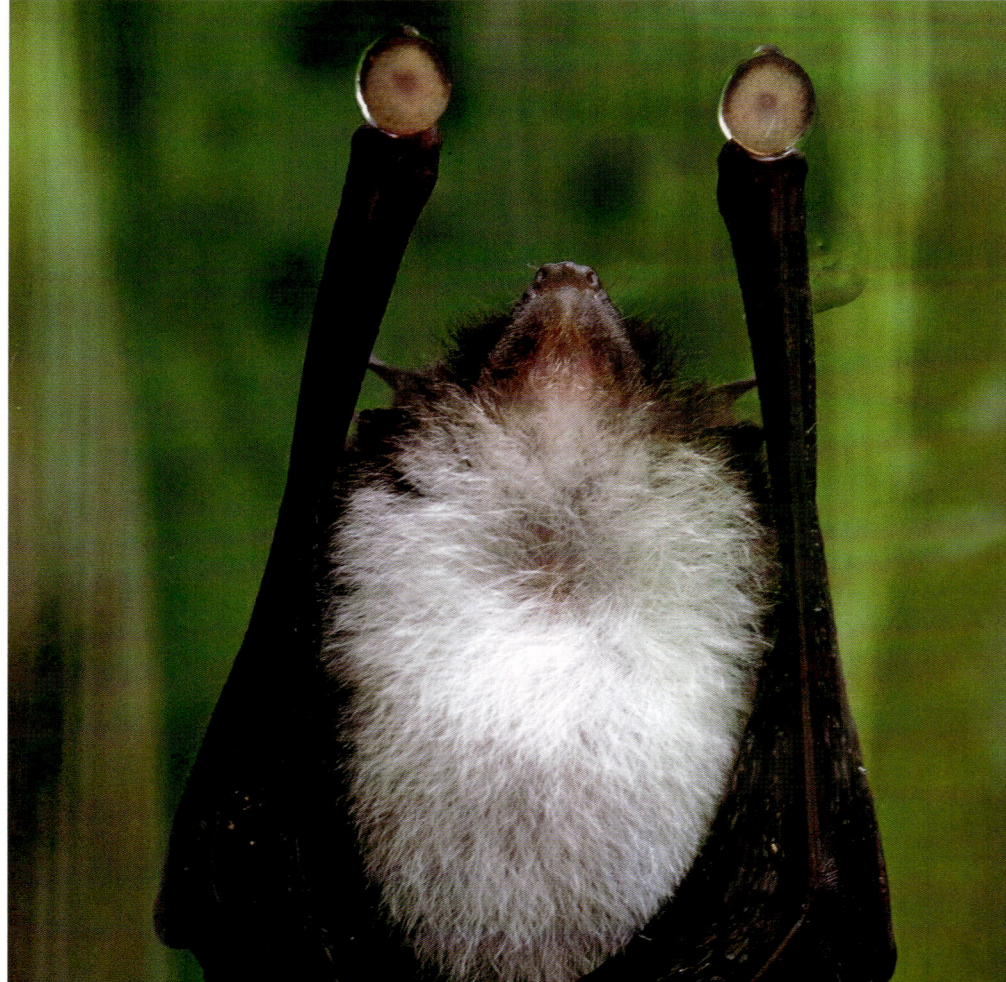

TOP In spite of its scientific name, the spectral bat (*Vampyrum spectrum*) is not a blood-thirsty vampire, but a predator of small vertebrates. It roosts in tree cavities on BCI. With its wingspan approaching 3 feet (1 meter), this impressive animal is the largest bat in the Americas. Photo © Marco Tschapka. **BOTTOM** A Spix's disk-winged bat (*Thyroptera tricolor*) adheres to the side of its roost using the suction pads at the base of its thumbs and on its ankles.

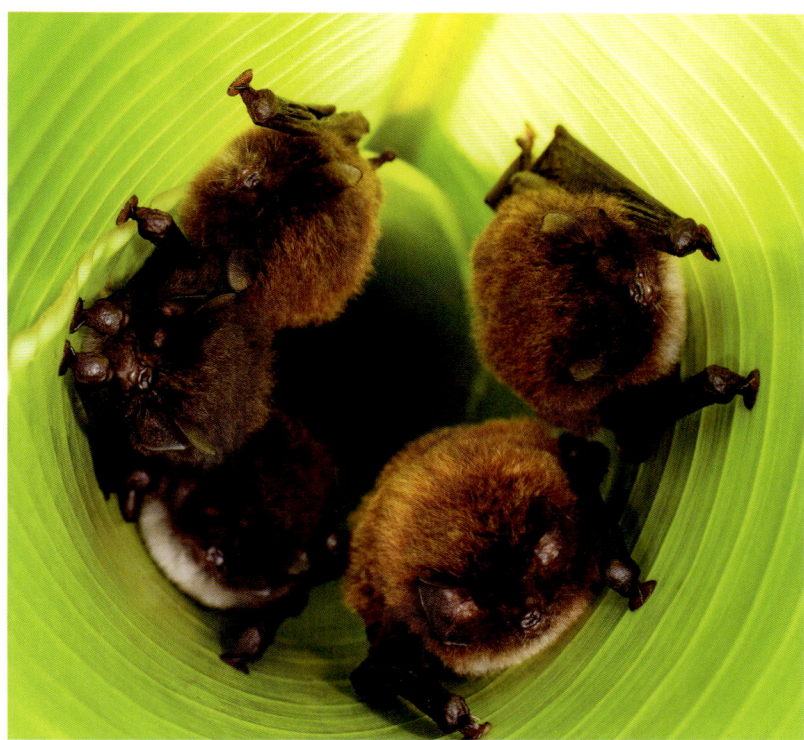

Not all BCI bats roost on parts of trees. One example of a unique alternative roosting location is in the young, furled leaves of the heliconia plant. One of BCI's tiniest bats, Spix's disk-winged bat (*Thyroptera tricolor*) has adapted so successfully to roosting inside these cone-like structures that it can no longer use any other roosts. These small 1-ounce (3-gram) bats use little suction pads at their wrists and ankles to hold onto the inside of the leaf cone—and when occasion arises, onto the fingernail of a curious bat researcher. Because the leaves unfurl quickly, the openings of these roosts are only suitable for the bats for about twenty-four hours, then the tubular structure of the unfurling leaf is lost, and the bats must find a new leaf while sticking together as a group. This task is made more difficult in BCI's dense forest because "heliconias like light, so they only grow at the edge of the forest and in treefall gaps. The bats have solved this problem by playing a game of Marco Polo, with group members that have found a new leaf of the correct diameter calling back and forth with the others until the full group is reunited in their new home.

Not all leaf roosts are so transient. On BCI, several species, such as Peters' tent-making bat (*Uroderma bilobatum*) and the pygmy fruit-eating bat (*Dermanura phaeotis*), construct tentlike roosts from leaves that can last for months. To make a tent, the bats use a palm frond, a heliconia leaf, or another boat-shaped leaf common in treefall gaps or at the edges of the forest. The leaf has to be high enough off the ground to be safe from ground predators, and the space under it has to be uncluttered enough so the bats can fly freely in and out. The bats then nibble on the leaf veins along the midrib causing the leaf to partly collapse and create a more protected space where they spend the day, usually in groups of females with their pups and a single harem male. Many of these species have white stripes on their faces or backs, which help to camouflage them, similar to the fur coloration in the greater sac-winged bat (*Saccopteryx bilineata*). Even though bats create their tents by chewing on the major veins of the leaf, the health of the plant seems relatively unaffected; leaf tents have been found to stay green and healthy for up to nine months.

While several species of bats on BCI construct tents from leaves, one species, the white-throated round-eared bat (*Lophostoma silvicolum*), is the clear winner when it comes to roost construction. Males of this species use their teeth to excavate cavities in active termite nests, which, apart from a brittle outer layer, are as hard as wood. *L. silvicolum* only occupies active termite nests. The presence of the termites makes these roosts safe, parasite free, and warmer than outside temperatures. Even though the cavity maintains the same shape if the termites die or abandon the nest, in the absence of the termites the cavity temperature fluctuates with outside temperatures, and *L. silvicolum* invariably abandons the roost if the termites leave.

A group of Spix's disk-winged bats (*Thyroptera tricolor*) roost together adhering to the smooth surface inside a leaf cone with their suction pads. This leaf will unfurl over the next twenty-four hours and force this group of bats to find a new roost together. The two nearly full-grown pups on the left can be distinguished from adults by their darker brown, less reddish color.

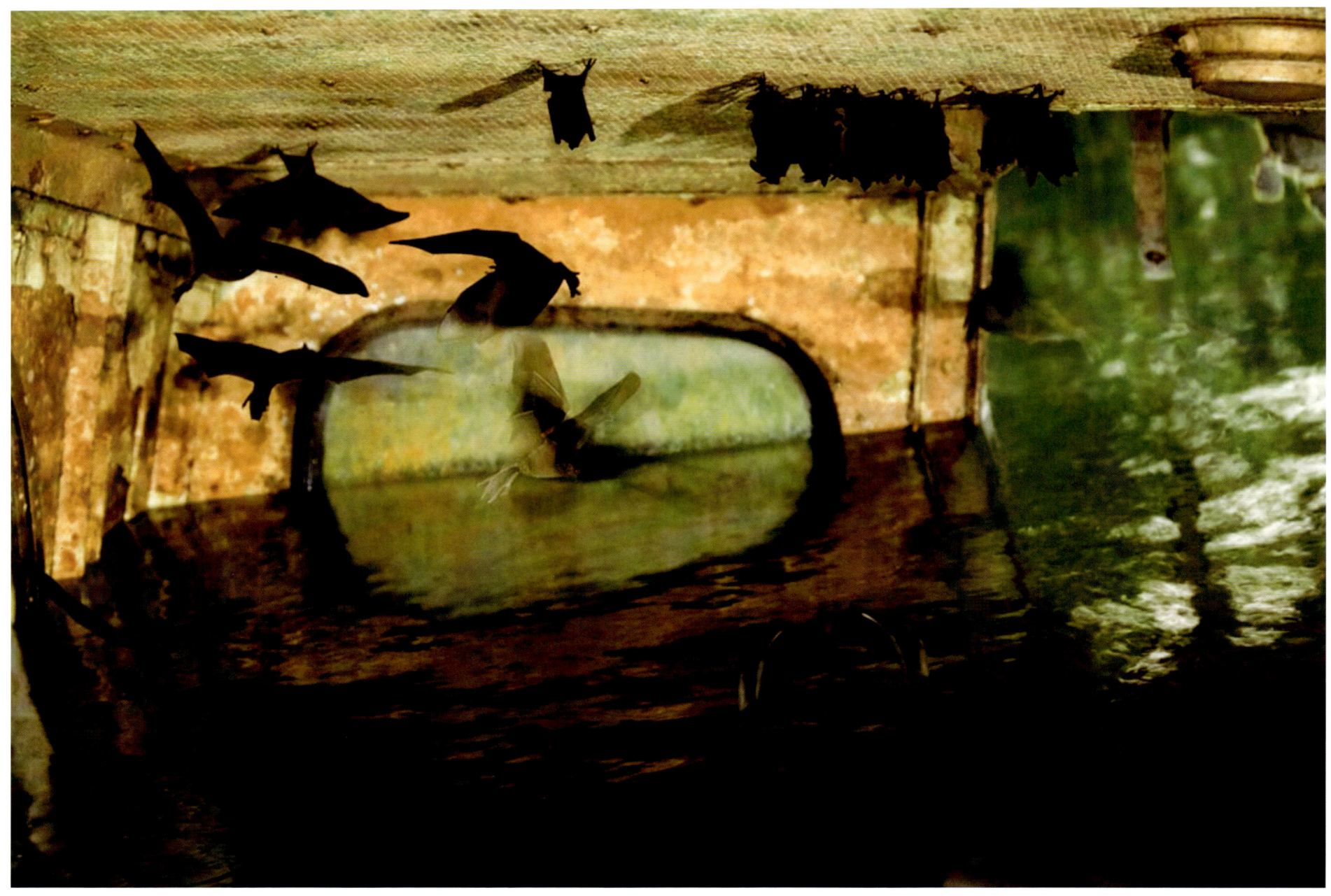

Bats take advantage of a wide range of safe roosting spaces, including this partially sunken boat that is home to a group of long-legged bats (*Macrophyllum macrophyllum*). From this roost, these bats emerge at night straight into their foraging habitat, over the water of BCI's Laboratory Cove.

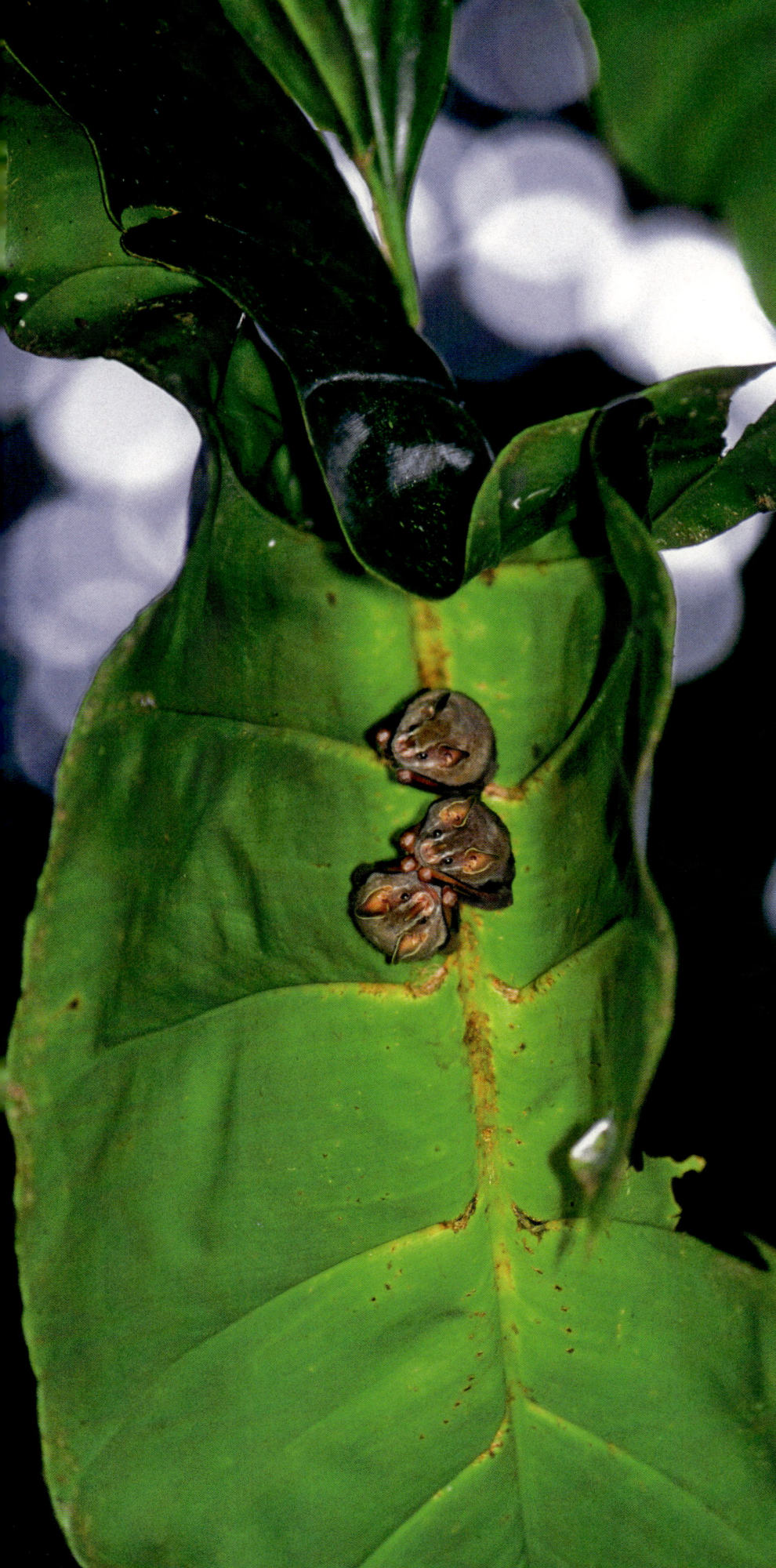

Research on BCI showed that single *L. silvicolum* males cut short their foraging time at night to patiently excavate and then maintain these roosting cavities, which can take up as much as half the volume of the termite nest. Once a male completes roost construction, a group of females joins the male, who is rewarded for his hard work by fathering pups. Other bat species occasionally take advantage of *L. silvicolum*'s diligent roost making. One of the largest bats on BCI, the greater spear-nosed bat (*Phyllostomus hastatus*), sometimes joins *L. silvicolum* in these roosting cavities. Since the omnivorous *P. hastatus* is not averse to occasionally including smaller bats in its diet, *L. silvicolum* suffers a risk when sharing their home with this intruder and potential predator. Although video recordings show that both adult and pup *L. silvicolum* are disturbed by these intruders, a predation event inside the roosting cavity has never been observed. It seems that it is usually young *P. hastatus* in search of a permanent group that temporarily squat in *L. silvicolum* roosts.

While some bats are very selective in their roost choices and are constrained by the availability of just the right roost, other bat species are much more flexible. Indeed, several bat species on BCI even take advantage of human constructions, including Pallas's mastiff bat (*Molossus molossus*) and the lesser bulldog bat (*Noctilio albiventris*), which roost in the lab buildings' roofs; Seba's short-tailed bat (*Carollia perspicillata*), which uses the island's lighthouses; the greater sac-winged bat (*Saccopteryx bilineata*), whose harems live on outside walls; and the long-legged bat (*Macrophyllum macrophyllum*), which roosts in a large colony in the cabin of a half-submerged boat in Laboratory Cove.

Several species of bats on BCI roost in tents they construct from leaves. These shelters protect them from the elements and provide an early warning system, alerting them to predator attack. An approaching snake or monkey, for example, will inadvertently vibrate the leaf, giving the bats a head start to fly safely away.

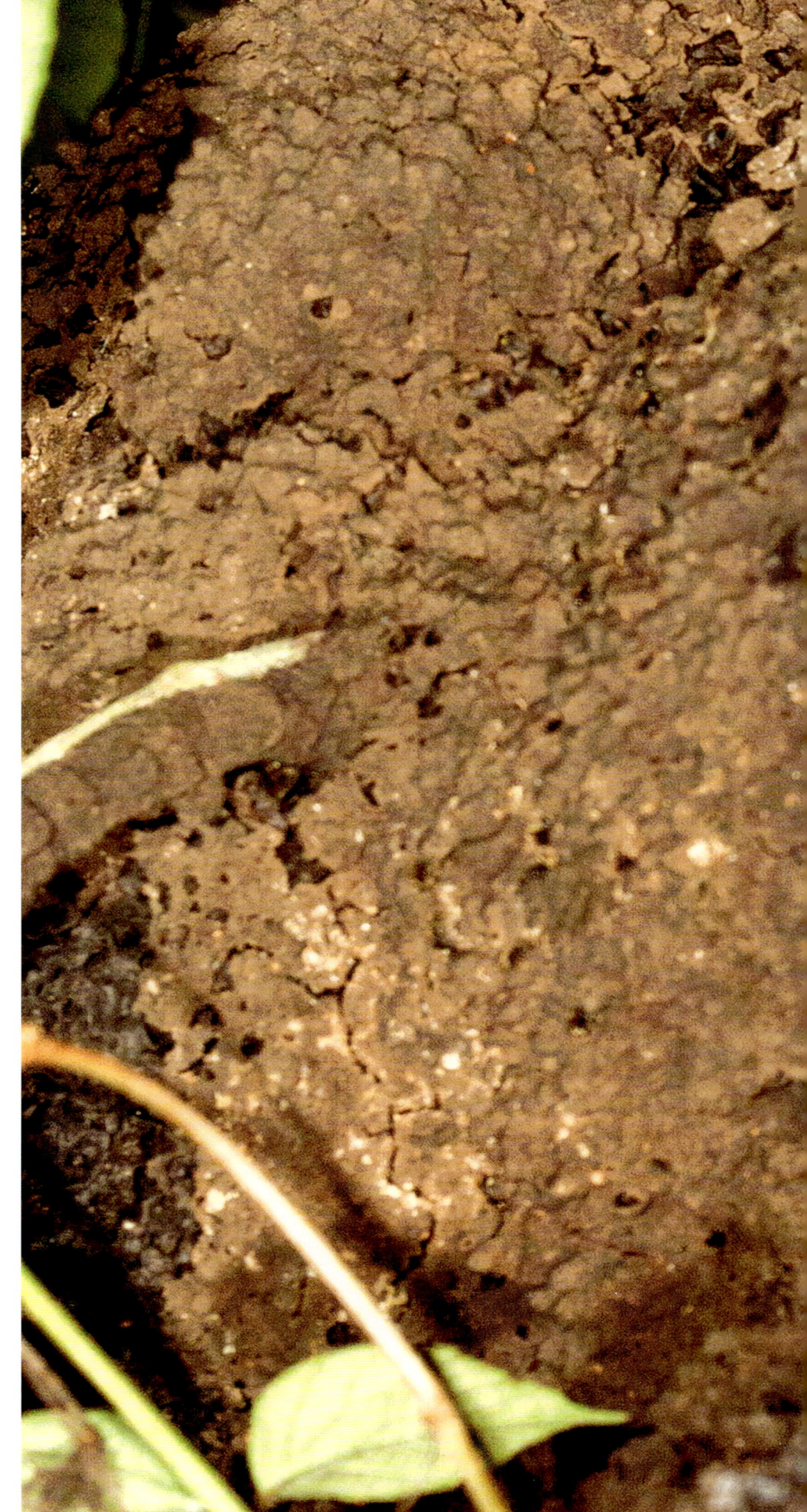

A close look at white-throated round-eared bats (*Lophostoma silvicolum*) in their termite nest. Young round-eared bat babies are born and remain naked until just before they are ready to fly. It is thought that this prolonged period without fur helps them minimize ectoparasite loads. The pups are able to remain naked thanks to the warm roosts they inhabit. The activity of the termites surrounding the bat cavity keeps roost temperatures stable and high.

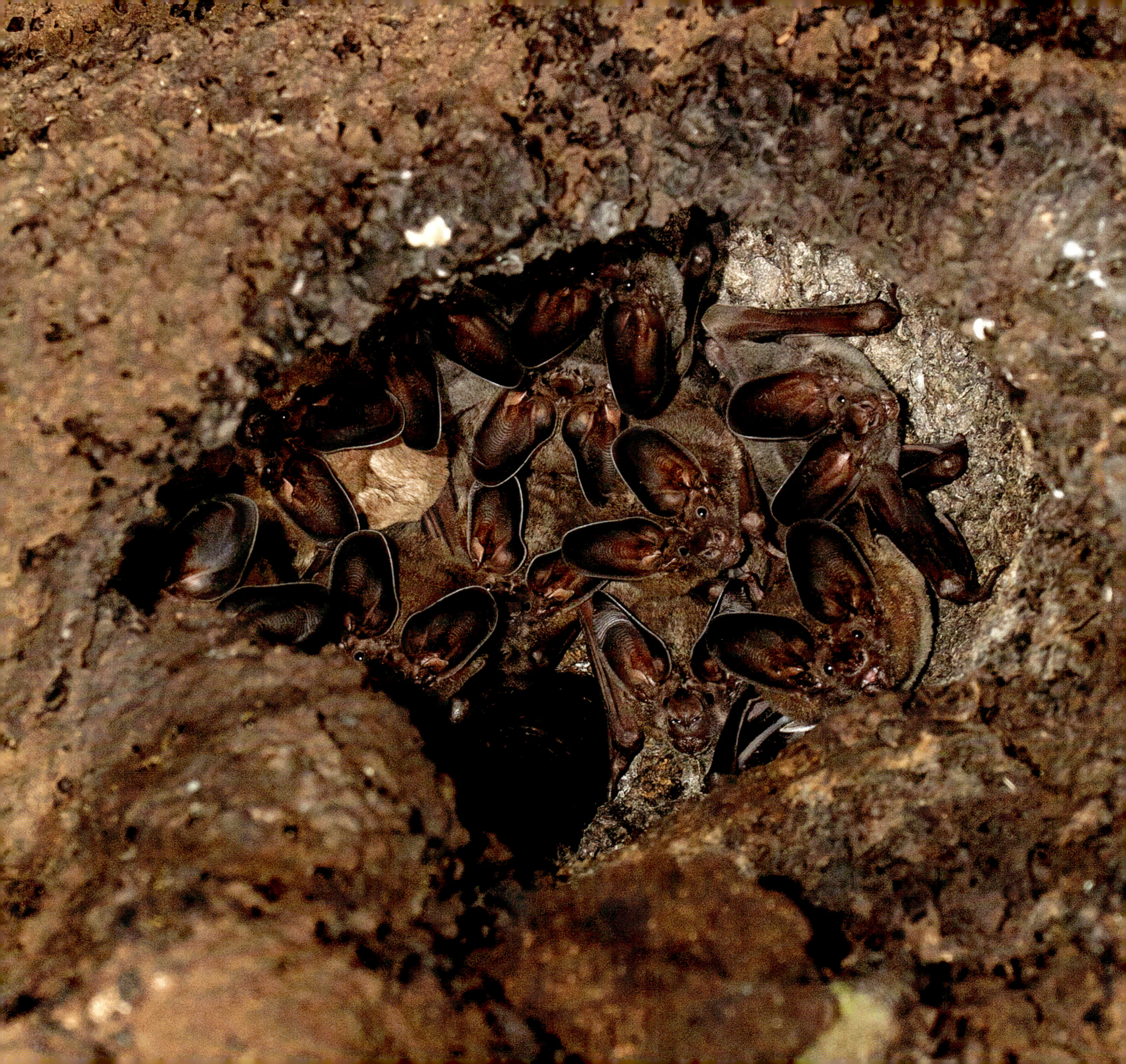

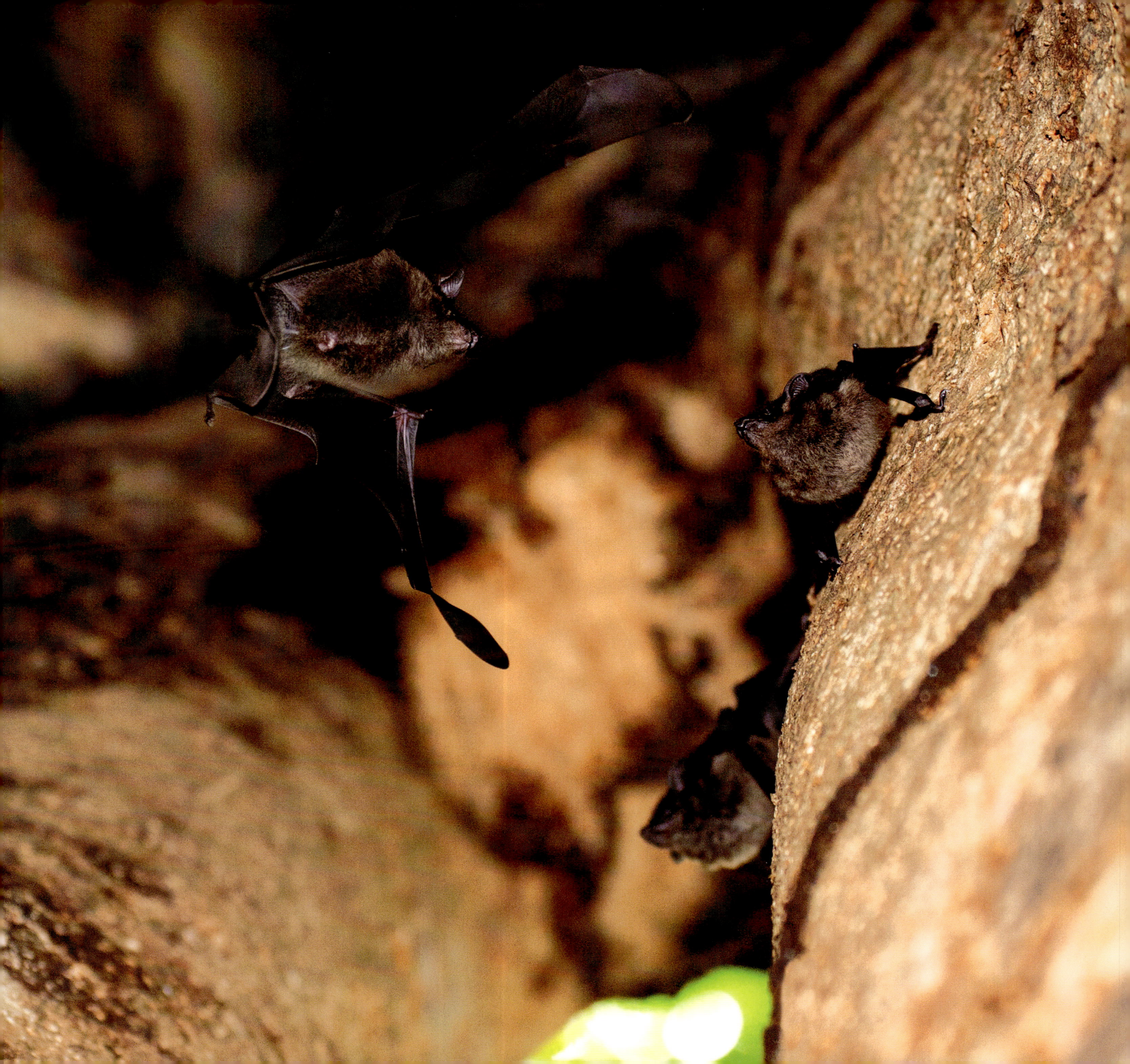

CHAPTER ELEVEN

LIVING TOGETHER, LIVING APART

SOCIALITY AND REPRODUCTION

Animals form many types of social groups to take advantage of the benefits of living with others. A social group can help group members find food faster and more reliably, alert them to the presence of predators, and provide warmth. Social groups in mammals span a wide variety of configurations, including loose aggregations of many individuals, tight-knit social groups dominated by family groups, a single monogamous pair and their dependent offspring, or a seemingly chaotic mash-up of males, females, and young, along with many other variations.

The outward appearance of these groups can be deceiving. Male-centered harems are actually formed by groups of unrelated females who raise their young together, maintain strong group bonds, and remain together sometimes for over a decade, allowing chosen males to temporarily live in their group. Large and apparently random aggregations of animals can contain many small subsets composed of individuals with close relationships to each other. And on the flip side, sometimes what appear to be social groups of bats are not social groups at all; they are disconnected individuals that happen to share the same roost.

Main drivers for bat sociality include benefits in finding food or raising offspring, and oftentimes both. Although having a reliable social group can be a considerable help, sociality comes also with costs. Animals may compete for food resources or for high-quality mates, lowering an individual's chance of survival or future reproductive success. Groupmates may also bring additional stress, lowered immune function, or infections and parasites. For sociality to evolve, these costs must be outweighed by benefits. For example, if finding food is the advantage sociality brings, food resources must be shareable and abundant enough to outweigh the cost of competition.

The complexity of a bat species' social system can be reflected in the complexity of its communication. Not only do male greater sac-winged bats (*Saccopteryx bilineata*) perform conspicuous courtship flights, wafting an odiferous cologne at females, but they also have a rich vocal repertoire, and sing for mate attraction and territorial defense just like birds do. Fascinatingly, *S. bilineata* is the only non-human mammal we know of to date that learns its vocalizations through babbling and imitation like human infants.

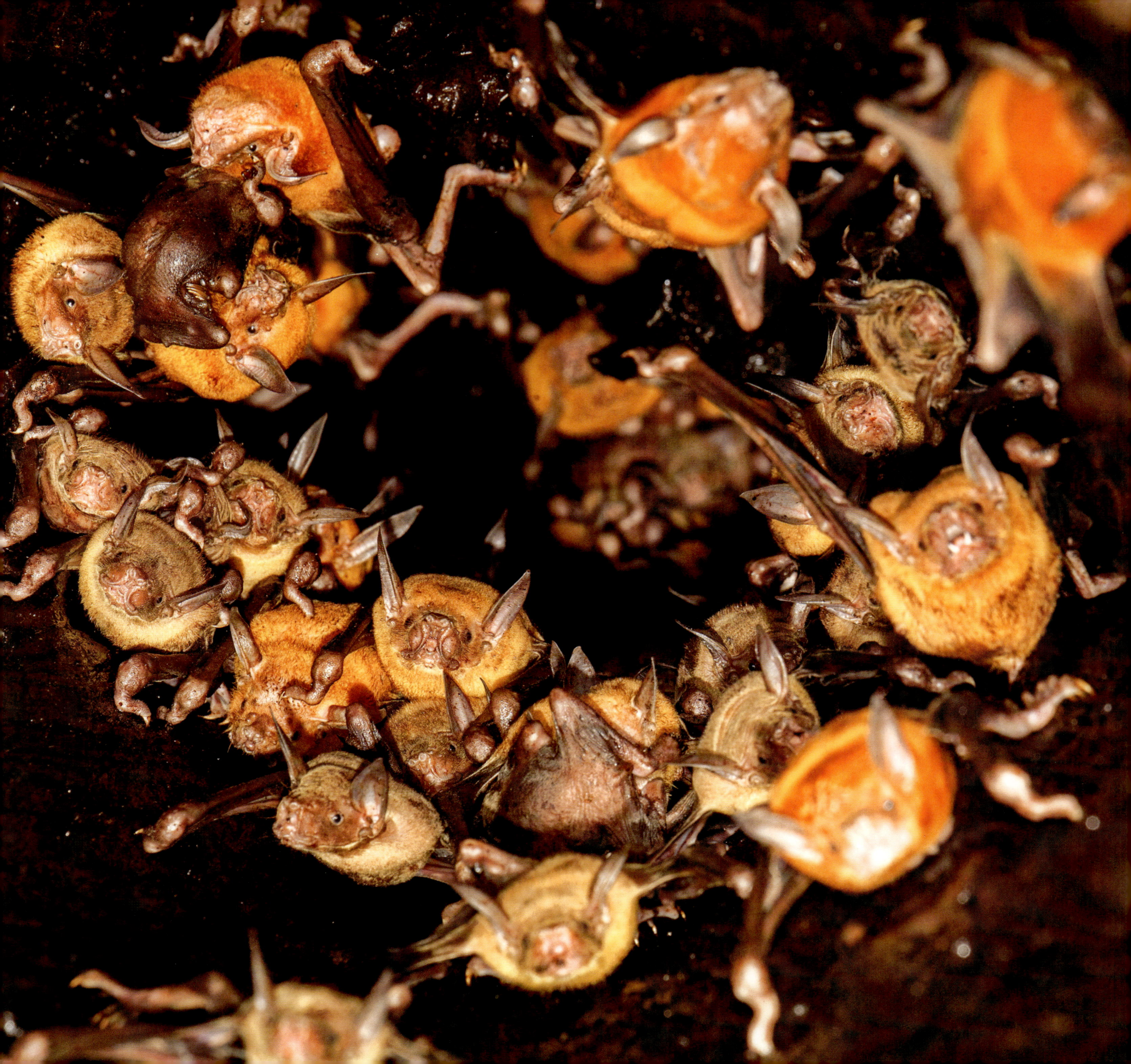

FINDING FOOD

Bats are experts at using social information to make decisions about when, where, and on what to feed. They eavesdrop on the echolocation calls of other bats as they feed on insects and fly toward the high-repetition feeding buzzes that signal a bat has found an insect swarm and is going in for the capture. To bats, these feeding buzzes are like opening a noisy bag of snacks in a quiet movie theater—everyone knows what is happening and the sound attracts those looking to share. Some species, like the lesser bulldog bat (*Noctilio albiventris*) and Pallas's mastiff bat (*Molossus molossus*), will forage with their groupmates and eavesdrop on successful foragers. These bats can recognize groupmates by their echolocation calls. They fly within hearing range of one another so that when one detects a swarm of insects all are quickly informed. Once a female *M. molossus* joins a social group she will not leave again. Groups of unrelated females who develop strong social relationships are common among tropical bats. Males come and go from these groups, but females are the stable backbone of most of these bat societies.

Here is a group of greater bulldog bats (*Noctilio leporinus*) with offspring of varying ages including several that had been recently born and are still naked. This and several other bat species are naturally a muted brown, but their fur can change to a deep orange color if the ammonia content from their accumulating urine in their roosts is sufficiently high. *N. leporinus* can be seen flying in small groups over the water in the evening.

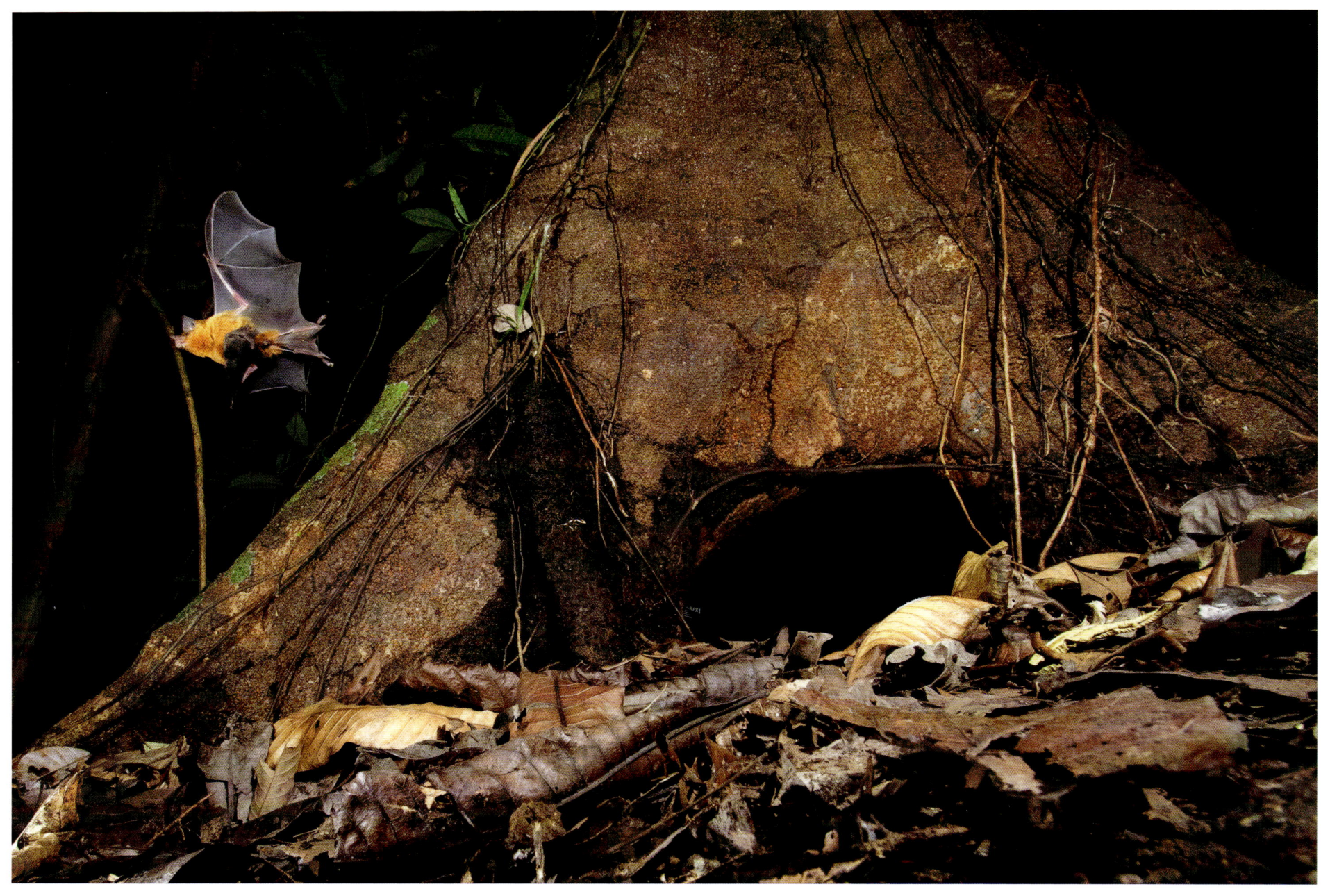

Many bat mothers will bring their infants with them while they forage, which can increase their flying load by 30 percent of their body mass or more. A Niceforo's big-eared bat (*Trinycteris nicefori*) leaves her day roost with her gray infant attached to her belly.

As we learned in chapter 6, the common vampire bat (*Desmodus rotundus*) is one of only three species of vampire bats, all of which feed exclusively on blood. *D. rotundus* is a true gourmet and highly specialized to feed on this single food source—an evolutionary strategy that comes with significant challenges. Blood is a low-quality food, containing a lot of water and relatively little nutritional content. Added to this is the complication that the bats' hosts are constantly on the move. However, once a large mammal like a tapir or cow is located, multiple vampire bats are able to feed on it simultaneously. If a vampire bat goes more than one night without feeding, it can quickly starve to death. On unsuccessful foraging nights, vampire bats rely on their family and friends to regurgitate blood for them. Hungry bats lick the partially digested blood directly from the donor's mouth. Interestingly, there is often competition to be the individual who *shares* blood with others; fed bats will beg to share their blood with hungry bats. By sharing with others, vampire bats cultivate critical social relationships that ensure their future survival on those unlucky nights when they cannot find their own meal.

RAISING OFFSPRING

Bats spend extraordinary amounts of time in their roosts with their roostmates, and in some cases this has resulted in incredibly cooperative relationships, such as those of the food-sharing vampire bats. In greater spear-nosed bats (*Phyllostomus hastatus*), unrelated females live together for their entire lifespans of up to fifteen years. These females develop group-specific social calls that help identify social group members across long distances at night. These screech calls take several months to learn. When a new female joins a group, all of the members adjust their group call to accommodate the newcomer. Females in these social groups synchronize their births and actively defend the pups of the entire group from attack by owls and other bats. Synchronized births among groupmates are common for many bat species, as there is enormous pressure to wean offspring by the time in the season that the maximum amount of food is available.

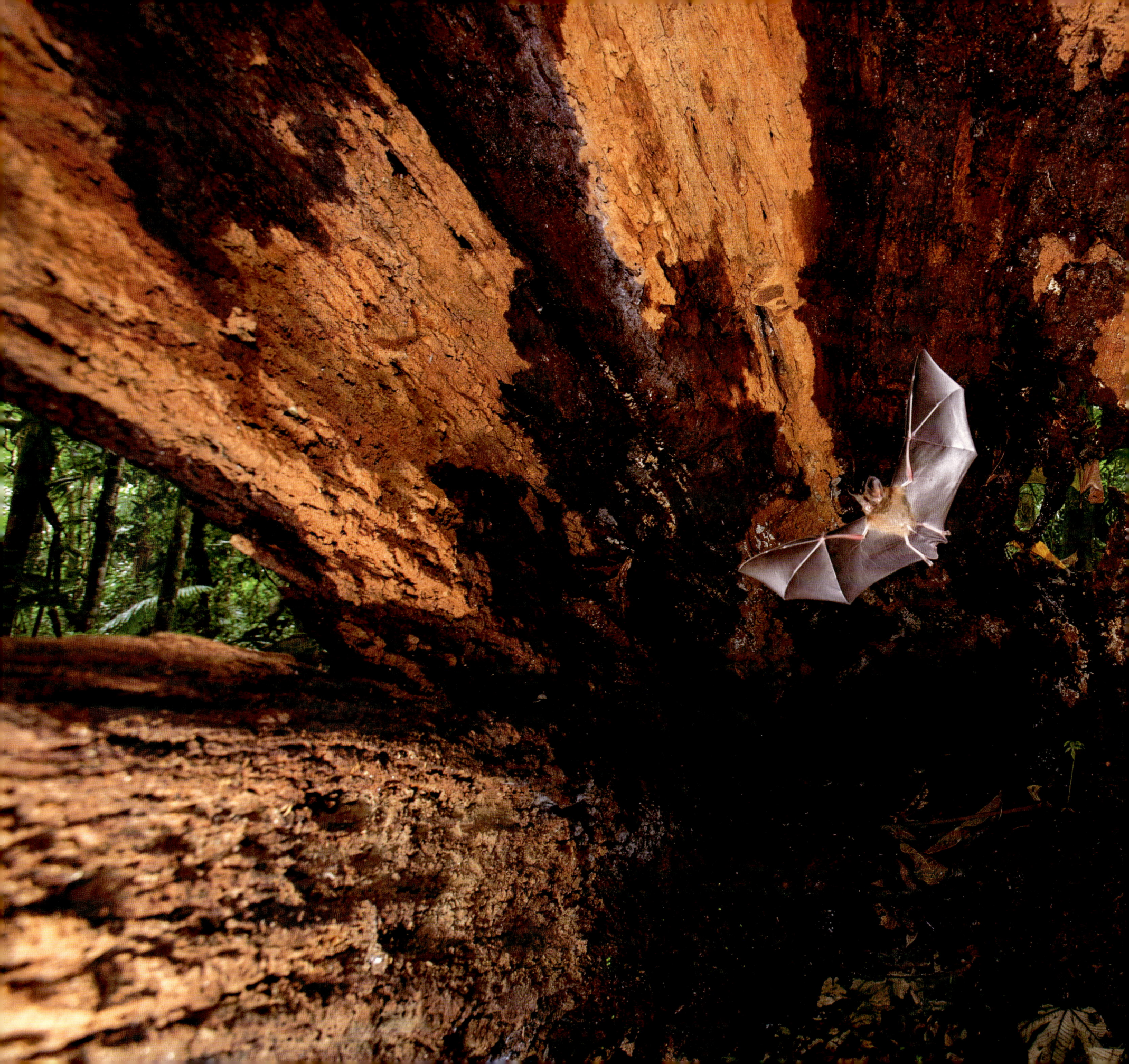

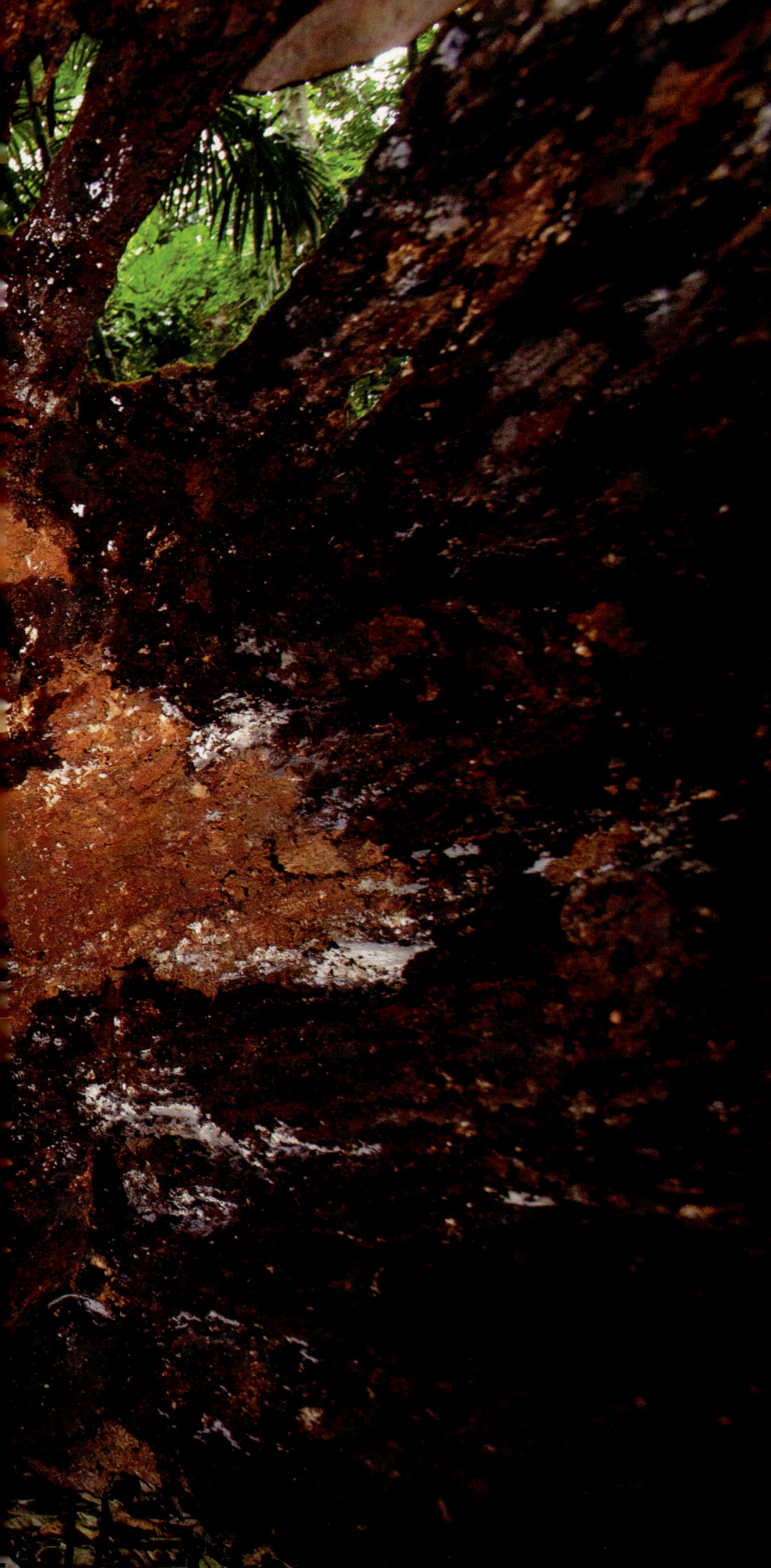

A view at dusk up into a big, hollow tree shows a fringe-lipped bat (*Trachops cirrhosus*) descending to exit its roost. At the beginning of the night, bats will often repeatedly dart in and out of their roost to check the situation outside. Only when it is dark enough, and thus safe from visually orienting predators, will they finally leave to forage.

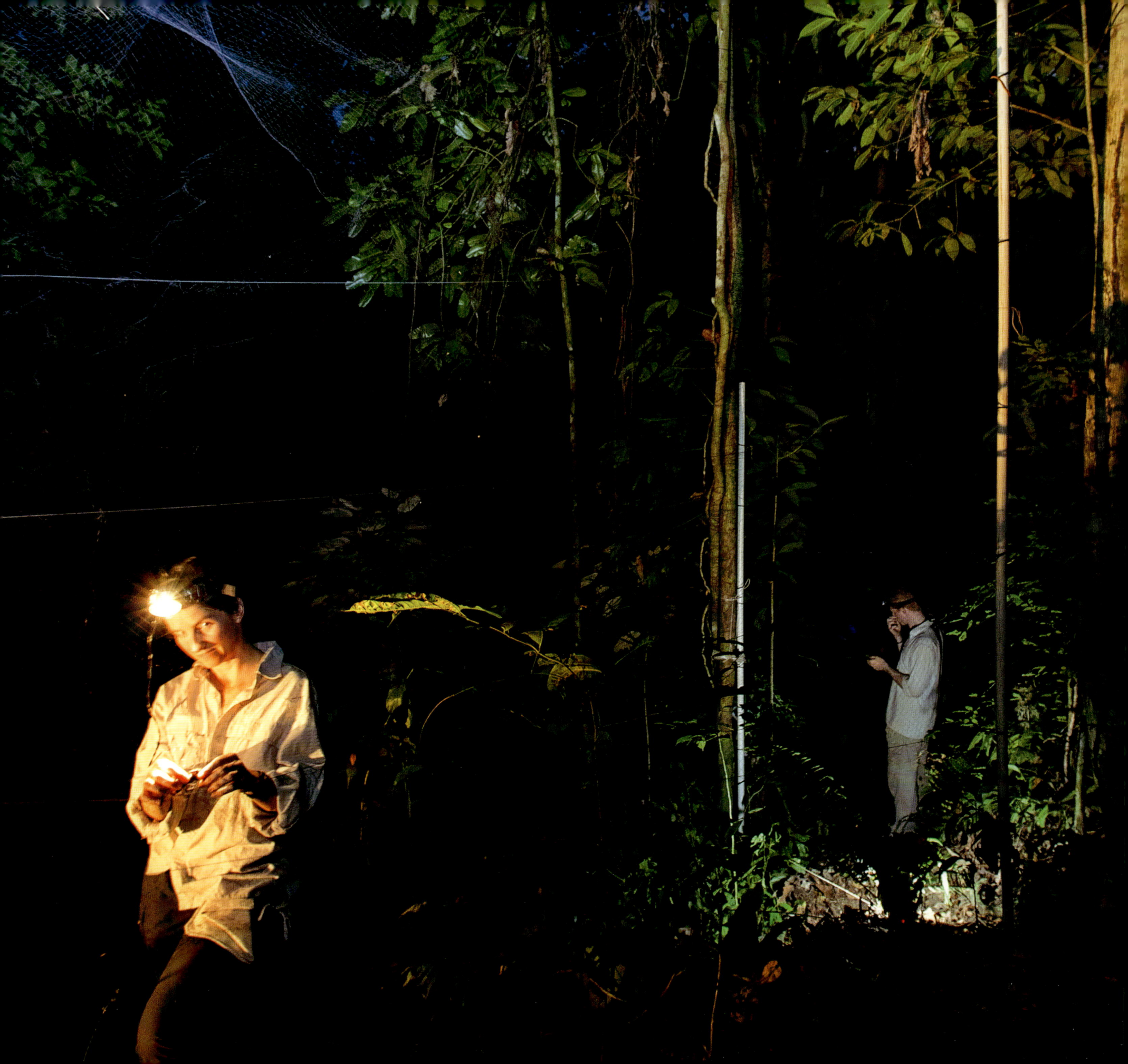

CHAPTER TWELVE

SHEDDING LIGHT ON THE DARK

NEW TECHNOLOGIES TO STUDY BAT BEHAVIOR

As we have learned, most bats are small and night active. They hide in dark caves, crevices, and cavities, and use sounds that humans cannot hear to perceive the world around them. These characteristics made it nearly impossible to study their behavior until technological breakthroughs in high-frequency audio recording and miniaturized radio tags allowed researchers to peer into their world. Before the availability of these technologies, scientists relied on classical zoological methods: Capture bats using mist nets, examine their diet by looking at droppings under a microscope, examine their morphology and physiology, and see how they behave in large flight cages. However, behavior in captivity is only a tiny glimpse of what bats are capable of. To really understand the diversity of bats it is necessary to study them in the wild. Technology has shed new light on old questions about bat behavior, biology, and ecology, and BCI has been a pioneer in offering the infrastructure and technical support to push the limits of innovation in tropical bat ecology.

Miniaturized radio technology was a fundamental breakthrough that offered an unprecedented window into the movement patterns of bats in the wild. Tiny radio transmitters, called radio tags, weighing less than 0.04 ounces (1 gram) that could be attached to a bat's back, allowed BCI scientists to track how individual bats used the island. Two-stage radio tags that, depending on the orientation of the tag, changed the speed of the sounds they emitted ("pings"), told researchers if the bat was flying (if the tag was horizontal) or if it was roosting upside down (if it was vertical). This very simple technology allowed researchers to assess the activity of bats by seeing how often and how long they flew. With advancing technology, other radio tags changed the frequency of their pings with bats' skin temperature, thus telling us about bat thermal physiology and adaptation to their environment, while even others allowed researchers to hear the individual heartbeats of a bat as it rested, flew, and found food. Hearing a bat's heart rate increase just before it flies off to find its first fig of the night is a dramatic window into its decision-making and energy expenditure.

Mist-netting is one of the key methods for capturing bats flying through the forest understory at night. Fine-meshed nets are strung between poles and anchored with ropes to stakes, trees, or other vegetation. Here, Smithsonian scientists untangle the bats that have become trapped in these forest nets.

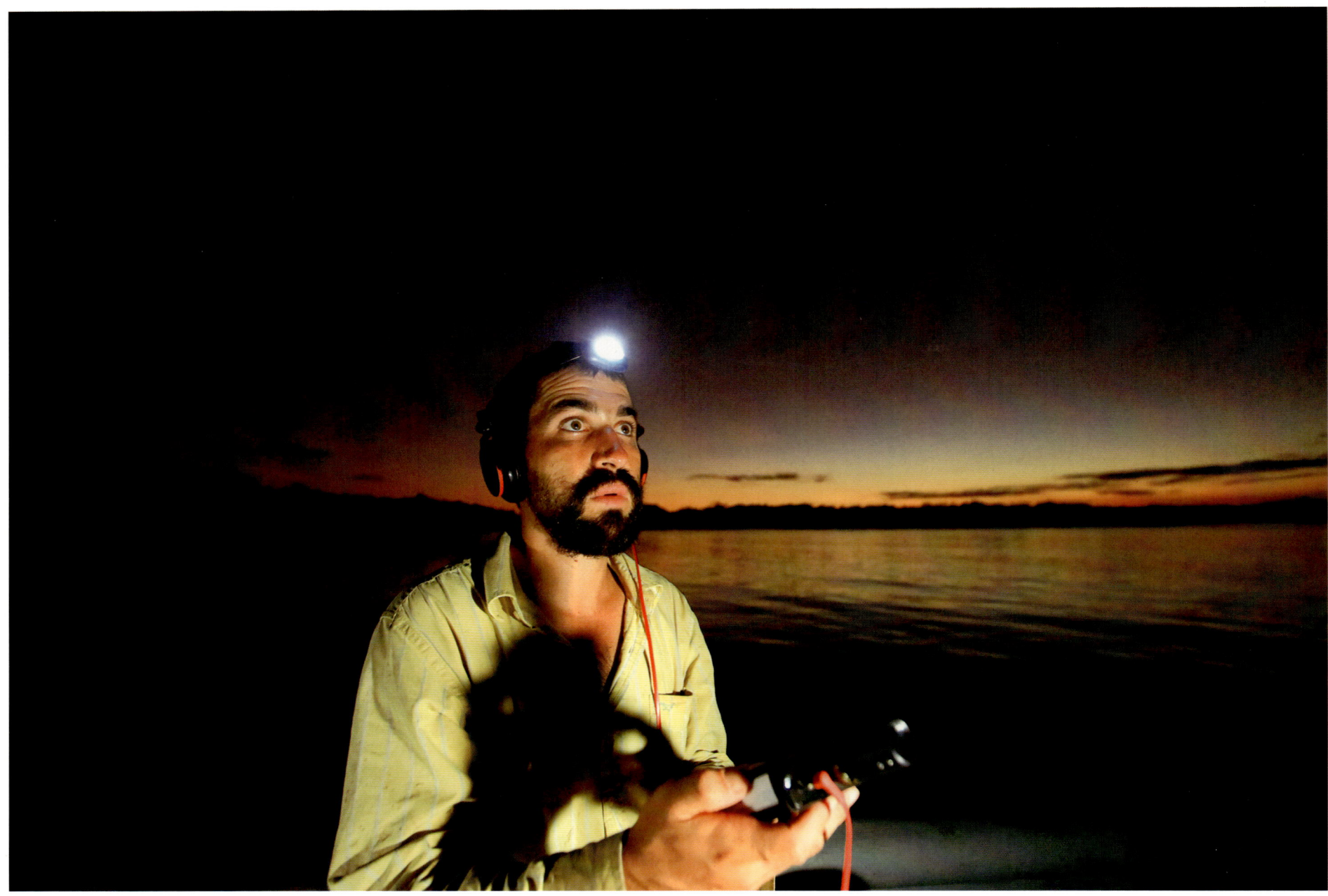

A Smithsonian researcher uses an ultrasonic microphone to detect and record bat echolocation calls. Many bat species have echolocation calls that can be used to identify them to species.

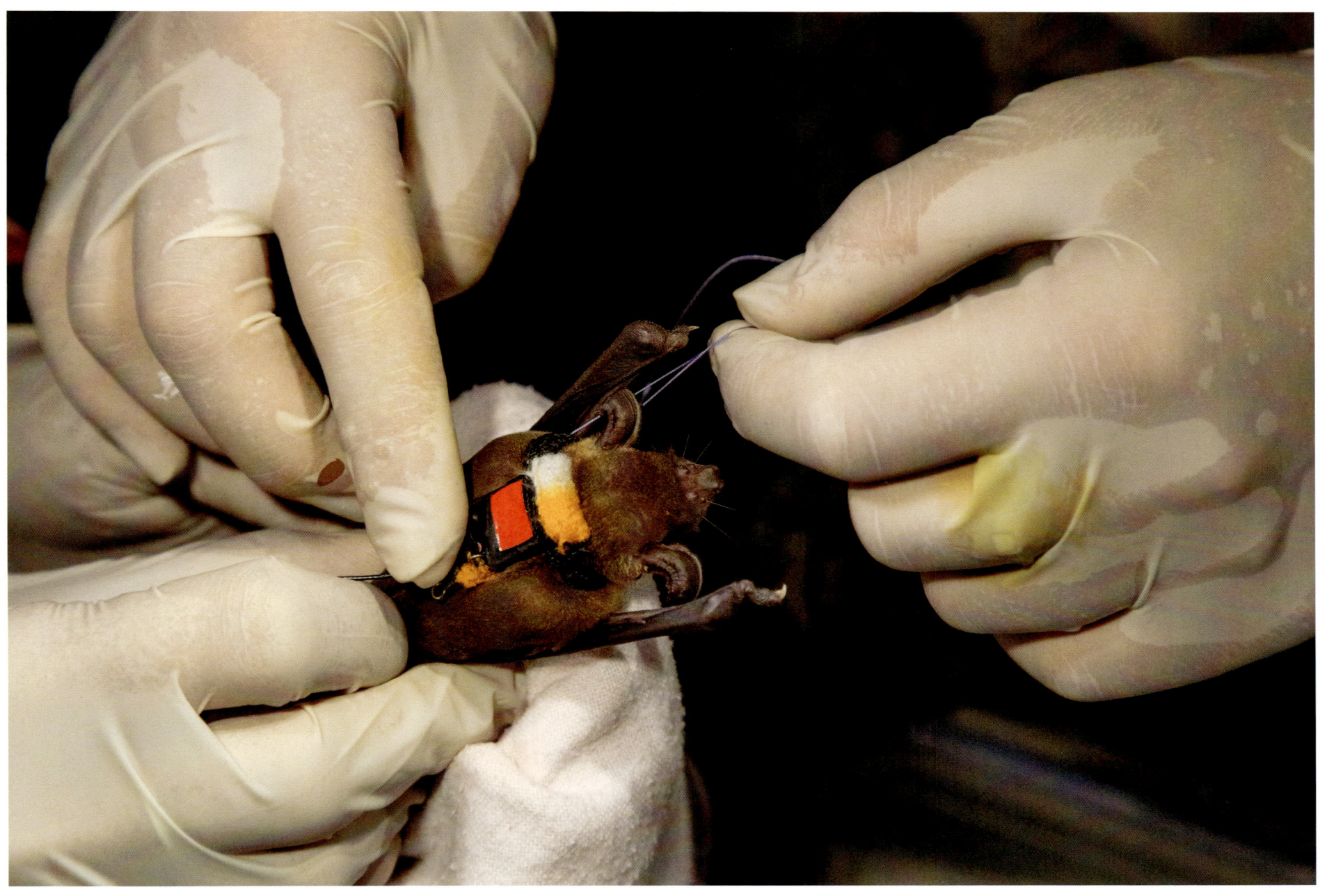

Miniaturized GPS and battery technology have allowed extraordinary insight into bat movement, habitat use, and social relationships. A greater bulldog bat (*Noctilio leporinus*) is fitted with a GPS logger on a collar. The collar is closed with purple biodegradable thread to ensure it falls off in the coming weeks.

Our ability to glimpse into the natural world of a bat increased enormously in the last decade with the advent of miniaturized advanced GPS technology coupled with additional sensors like accelerometers and barometers. These devices measure more than just simple spatial movements, giving new insights into how often bats flap their wings, how much energy they expend, and how high they fly. The data that can be gleaned through GPS devices is astounding. However, GPS is a power-hungry tool, and despite improvements in battery technology, power management, and data-transfer capabilities through communications like Bluetooth, the weight of current batteries make GPS tracking devices still too heavy for many BCI bats to carry.

The time and persistence of bat researchers can replace GPS technology to a certain extent, and scientists have taken advantage of BCI's excellent research infrastructure for decades to do so. Several flight cages in the forest have allowed researchers to bring bats into temporary captivity for close observation and detailed experiments to understand their behavior, sensory biology, and ecology.

The controlled environment allows the use of high-speed cameras synchronized with ultrasound recording equipment to help resolve bats' fast movements and echolocation. For example, by offering bats pools of water in flight cages and closely observing them, we now understand in much greater detail the foraging behavior of a bat capturing prey from the water surface. While these flight cages have to be serviced continuously, since they fall victim to termites, mold, and other ravages of a tropical climate, they are incredibly useful for many different types of experiments.

One of the first applications of infrared video cameras on BCI was to study the fascinating behavior of the common big-eared bat (*Micronycteris microtis*), which uses the stairways of several dorms for roosting. Being able to film this species night after night has been pivotal for the discovery of the maternal behavior of these bats. They not only feed on insects larger than themselves, but also bring prey to their nonvolant offspring (see chapter 4 on learning). Infrared cameras have now become part of the basic equipment of bat researchers in Panama and elsewhere.

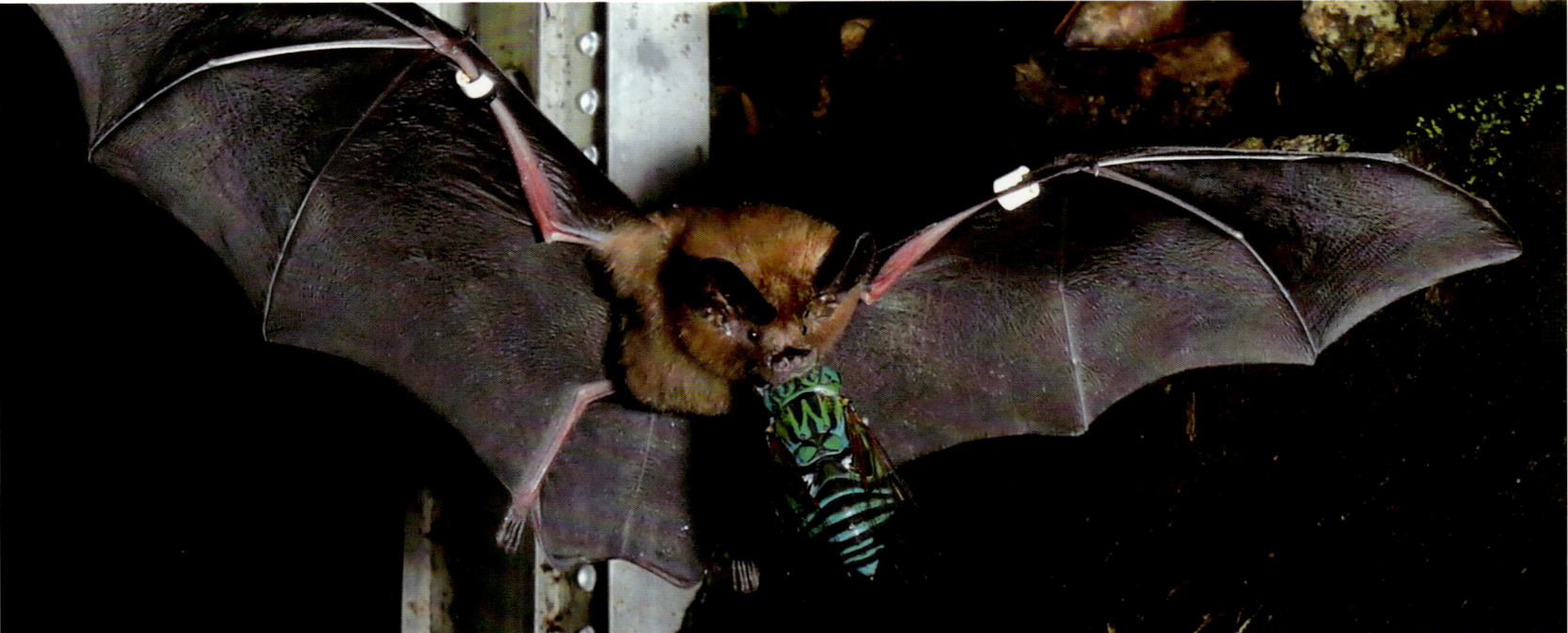

BCI was also one of the first places where transponder technology was used to monitor bats. Recognizing that some species, like the white-throated round-eared bat (*Lophostoma silvicolum*), did not respond well to external markings such as wing bands or collars, scientists started to mark them with subcutaneous passive integrated transponders (PIT tags). By equipping their roosts in termite nests with antennas powered by car batteries, scientists were able to closely monitor individual bats coming and going and assess group composition and stability. Over the years, PIT tags have been crucial for numerous long-term studies. For example, PIT tags helped to discover the shortest-known life span in a bat to date: an average of 1.7 years for Pallas's mastiff bat (*Molossus molossus*), a species which lives under the roof of BCI's lab building.

As emphasized in previous chapters, one of bats' key innovations is their ability to use high-frequency echolocation. It was only relatively recently—in the 1940s—that echolocation was recorded for the first time. Some of the first field recordings on BCI were made with Eli Kalko's delay line, constructed in the 1990s by the University of Tübingen's workshop in Germany. This custom-made device used digital memory for temporarily capturing ultrasonic bat calls, which were then slowed down and saved on a connected tape cassette recorder. Combined with a multiflash camera system, this device allowed Eli to explore how the greater bulldog bat (*Noctilio leporinus*) used its echolocation calls to capture fish as it flew low over the cove at BCI.

We have come a long way from the first ultrasound recorders that were used to record bats in the laboratory on cassette tapes. Echolocation recordings have now become fully digital and use ultrasound microphone interfaces that directly save bat vocalizations on a computer hard drive. Many clever experiments utilizing these tools have revealed the feats achieved by bats, including how the tiny common big-eared bat (*Micronycteris microtis*) can find insects sitting still on leaves, a fine-scale challenge for sound detection; how fringe-lipped bats (*Trachops cirrhosus*) can detect ripples on the water surface made by their túngara frog prey; and how baby greater sac-winged bats (*Saccopteryx bilineata*) use babbling—much like humans!—to learn social calls from their mothers.

The steady influx of new scientists and their students at STRI ensures the evolution of state-of-the art research. Many new technologies are put to the test at STRI, such as miniaturized digital audio recorders that can be attached to a bat's back and proximity loggers that track how close to each other bats fly during group foraging. In addition, modern molecular methods have unveiled many secrets hidden in the bat genome. Bat genetics help to understand phylogenetic relations among the species, and molecular tools can tell us whether a bat has been in contact with pathogens, and, if so, which ones. With ever evolving technologies at our fingertips and teams of top international and local researchers, BCI and Panama will remain hot spots for top-notch bat research for generations of scientists to come.

OPPOSITE Fine-scale monitoring of bat colonies in the wild is challenging, as many bat species roost in areas inaccessible to researchers and defy direct observation. One technology that has shed light on colony dynamics are light-barrier systems. Light beams are emitted across the entrance of a roost; as a bat flies through the roost entrance it disrupts the beams, triggering recording devices. Light-barrier systems can be coupled with cameras for visual monitoring and ultrasonic microphones to record echolocation calls. On BCI, Smithsonian researchers monitor a colony of common big eared bats (*Micronycteris microtis*) in a storm drain near the laboratory buildings. These bats leave to make short foraging trips at night, returning with prey to consume in the safety of the roost and to share with their pups. Here, a female carries a large cicada; the metal panel of light-emitting diodes can be seen behind her. Photo © Inga Geipel and Karl Kugelschafter.

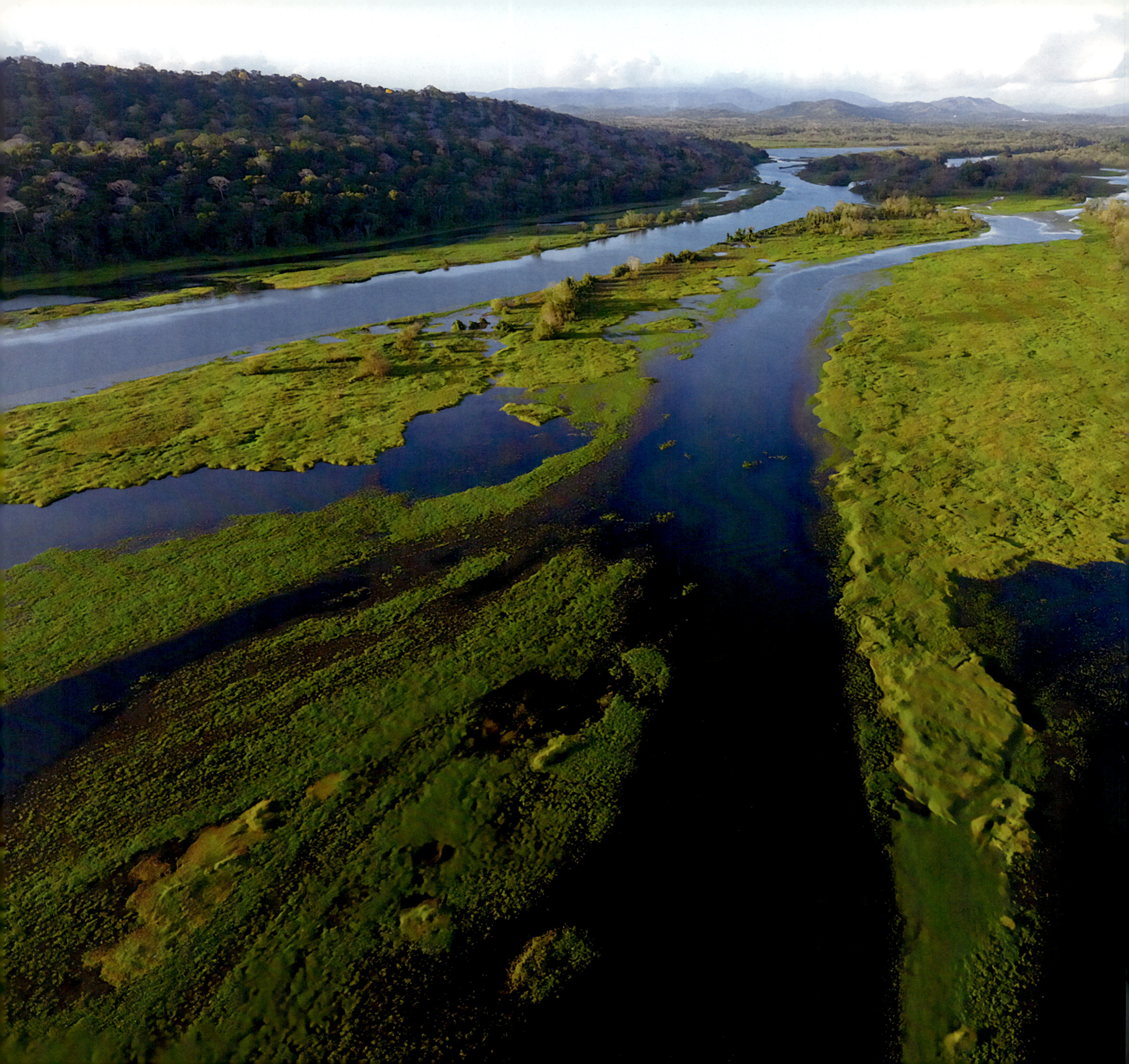

CHAPTER THIRTEEN

BATS FOREVER?

CONSERVING THEIR HIDDEN WORLD

Bats live nearly everywhere. They occur in wild areas all over the planet, including regions densely populated by humans. Mostly they are inconspicuous, going about their nightly activities undetected by the human eye. Yet, they are omnipresent, hunting insects, pollinating flowers, and dispersing seeds—activities we take for granted. With over 1,450 species, bats are diverse, and their conservation needs are similarly varied. Nearly a third of the bat species assessed by the International Union for the Conservation of Nature (IUCN) are ranked as threatened with extinction or data deficient.

Bat populations are declining across the world, and the top-ranked threat in the IUCN Red List for bat species is habitat loss, including land use changes such as agriculture and logging. In North America, the invasive fungus that causes White-Nose Syndrome has added to these threats, eliminating up to 99 percent of several bat species. The dramatic expansion of wind power that is needed to transition to carbon-free energy sources also poses a large risk to bats. Millions of bats are killed each year by wind turbines across the globe. Operating wind turbines above minimum wind speeds (curtailment) can decrease bat mortality, but comes at a loss of revenue for the facilities. As technologies develop, these green-on-green conflicts will need to be resolved to maintain climate goals and keep bat populations healthy.

Complex landscapes with intact forests and clean water are essential for bat health and maintaining future populations. These habitats provide a wide range of food and roosting options for tropical bats.

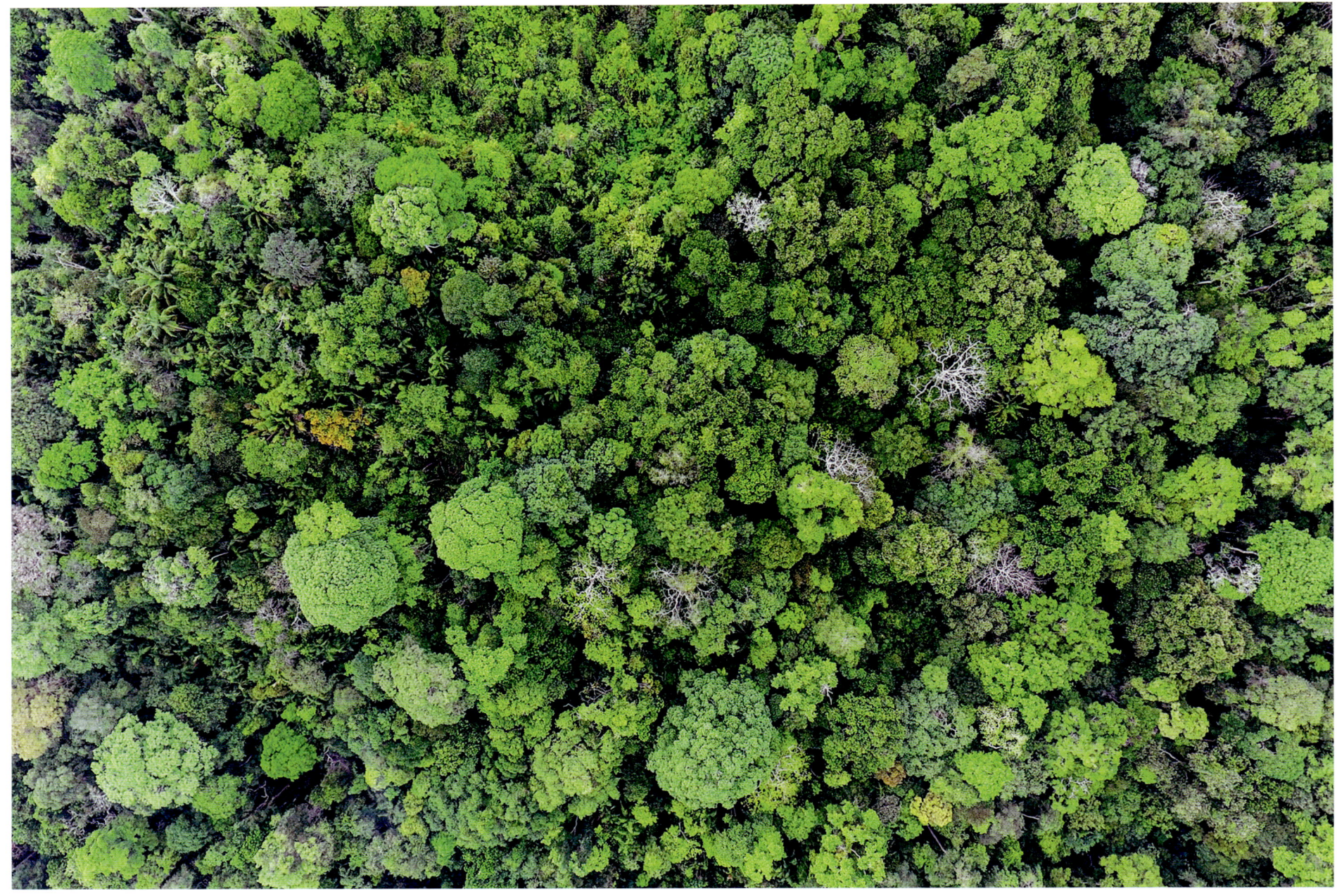

Tropical forests are incredibly diverse and provide resiliency against the effects of human-induced climate change. The ongoing loss of diverse forests destroys habitats essential to a wide range of species.

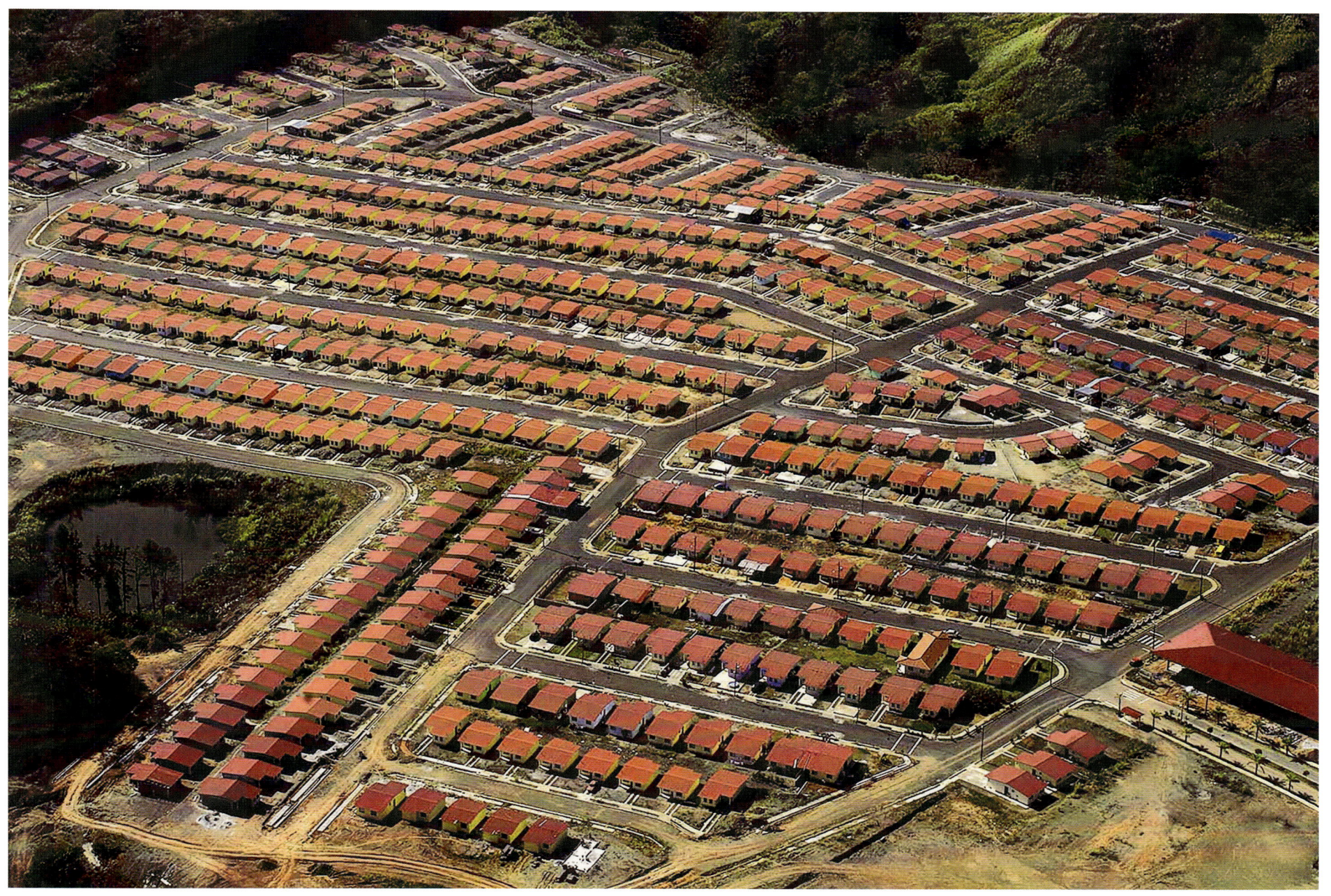

The rise in human populations puts enormous pressure on forested resources, and accessible landscapes are increasingly claimed for agriculture, logging, and housing.

The conservation threats to bats are all compounded by their cryptic lives. We know little about the population status of most bat species, particularly those in the Global South, which makes prioritizing and planning conservation actions challenging. Bats need places to eat and to sleep and depend on healthy ecosystems to provide food and undisturbed roosting options. Unfortunately, across much of the planet pristine areas are becoming increasingly rare. Most bats feed on insects, and global declines in insect availability threaten the viability of many bat species. This insect apocalypse is likely caused by a combination of factors including habitat loss, pesticide use, climate change, and increasing light pollution, all a result of an ever-growing human population generally overshadowing the basic needs of every other organism. Rainforests are disappearing at an unprecedented rate. We radically reshape entire landscapes according to our needs for food, and an increasing part of our world is being converted into treeless concrete deserts. Formerly continuous stretches of forest are being split into smaller chunks that are isolated from each other, and many bat species cannot cross the open areas between forest fragments. Bat foraging opportunities are disappearing, and roosting sites are often no longer available.

Some bats, however, can make use of human-dominated habitats. Bats can be regularly spotted in gardens, parks, and even in the middle of cities. But there is considerable variation in bat adaptability to human-modified landscapes. There is a small subset of species that will happily accept any crevice in a house as a roost, while others will only inhabit roosts they made themselves from the leaves of specific plants. Our human-modified landscapes will continue to create some winners but also many losers in this unfortunate habitat sweepstakes unless specific conservation actions are taken to protect bat habitats.

Humans initially provided some useful additions to the landscape from the perspective of certain bats. Species that roosted in caves benefited from buildings. Under natural conditions these bats were restricted to rocky habitats, but human buildings

The complex life around Barro Colorado Island is fed by the Chagres River, seen here during the dry season with the yellow flower of guayacán trees (*Handroanthus guayacan*) dotting the landscape.

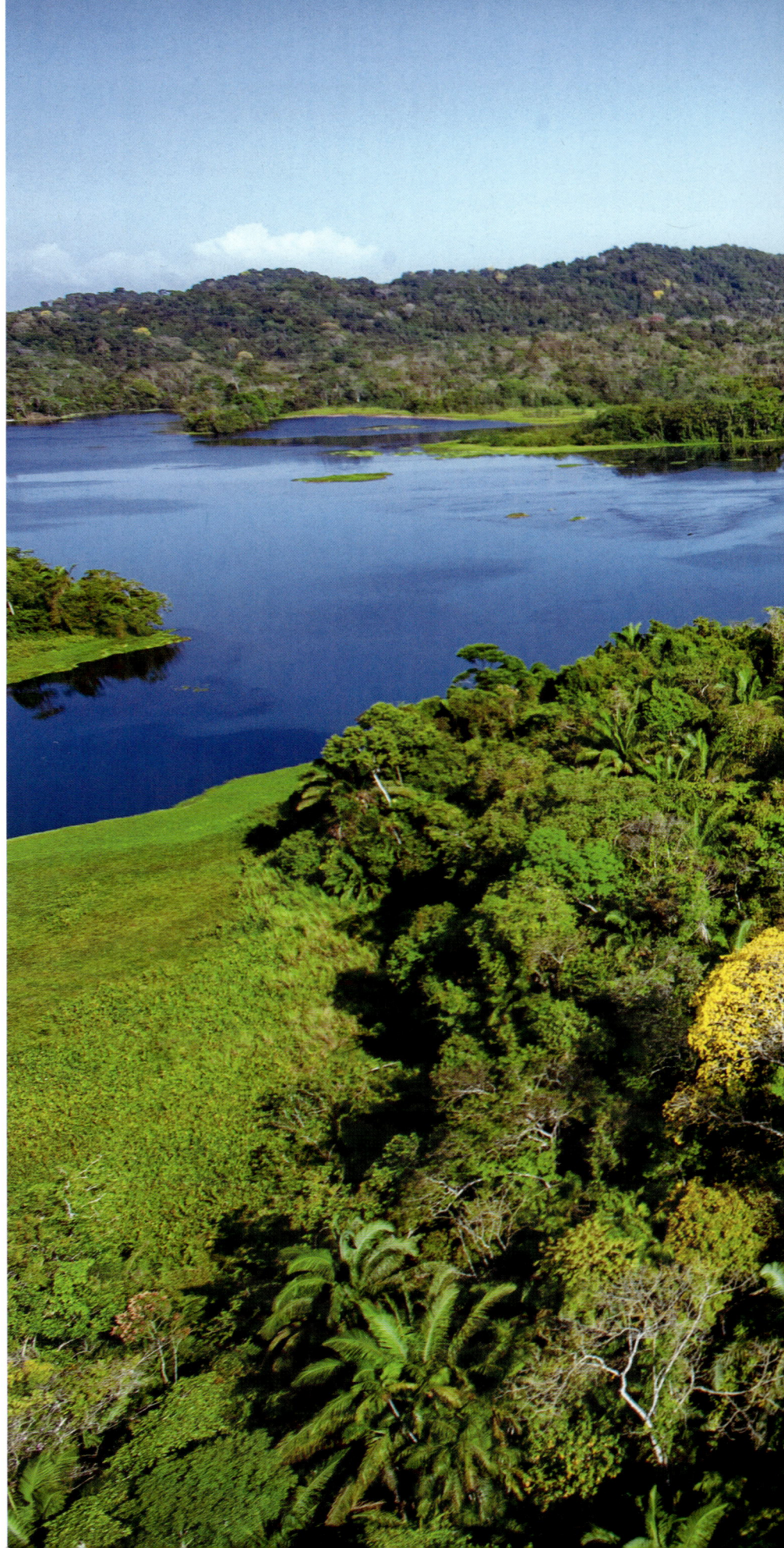

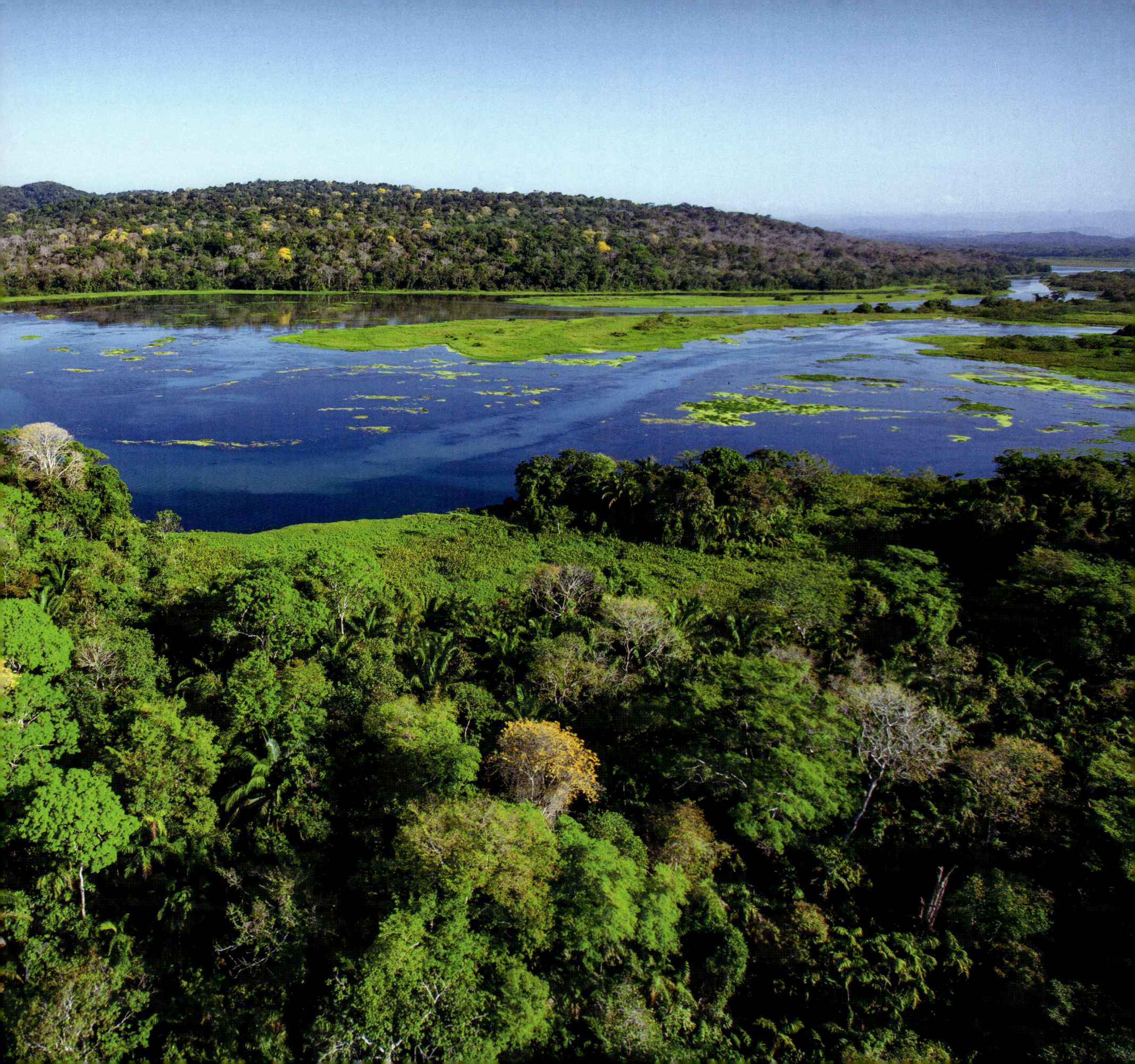

increased the availability of similar roosting options, and perhaps for some species even allowed an expansion beyond their original distribution range. Attics, where warm summer air is trapped, offered roosting options to bats looking for warmer roost sites to raise their young. Bat species that use crevices to roost found many new options in the cracks and cavities of human buildings as well as in abandoned mines.

However, some of these advantages have been disappearing. For instance, homes are now better sealed and insulated than in the past, reducing access to optimal roosting spaces. The proliferation of electric lighting offered advantages for some insectivorous bats, which quickly learned that the lighting attracts insects and creates dense patches of food. Consequently, bats frequently can be observed hunting around streetlights and floodlights of sports fields, but these are limited to a few generalist species; most bats tend to avoid artificial light at night. Light pollution is not only a major conservation challenge for most bat species, but also for other nocturnal creatures. Insects can be disoriented by artificial lights they mistake for the moon and stars, cues they typically use to orient. Once close to an artificial light source, they often die from colliding with the light, overheating, dehydration, or from predation by bats or other insectivores.

Climate change has altered the living conditions in many ecosystems and has intensified conservation challenges to bats. The extremes or shifts in weather that are a consequence of climate change result in drought, unprecedented freezing, or other disruptions that reduce food resources. Because bats seek out particular temperatures for roosting and raising their offspring, warming local temperatures reduce the number of usable roosts that are available. With warmer temperatures, bats are less likely to migrate and move their ecosystem services across their historical range. There may be increasing mismatch in the timing of when bats appear on a landscape or emerge from winter hibernation and the food they depend on. Changes in land use to support human populations removes environmental buffers that bats had historically depended on, putting large portions of ecosystems at risk.

The services that bats perform voluntarily for the ecosystem are highly valuable, but we often take them for granted. Sometimes it takes the disappearance of bats to highlight their ecological and economic value. For example, durian fruits in Southeast Asia are highly sought after and form a multimillion-dollar industry. The flowers are pollinated by fruit bats looking for floral nectar. As these bats become rarer in certain areas, some durian plantations have experienced a lack of pollinators, resulting in reduced crop sizes. Consequently, in these areas durian flowers now have to be pollinated through expensive manual labor using small brushes. Bats are also important pollinators of agaves, from which tequila is made. Insectivorous bats save agricultural production an estimated 23 billion dollars per year in crop damage and pesticide costs just in the United States—and provide an environmentally friendly solution to the overuse of pesticides in the bargain. Without bats, some of our most prized agricultural products could be more expensive to produce or no longer available.

So what can we do to protect bats? First of all, bats need friends. Because bats are mysterious creatures that forage in the dark of night, many people are afraid of them. The recent Covid-19 pandemic caused particular concern about the health risks associated with bats. However, there is still no support for the direct transmission of the SARS-CoV-2 coronavirus from bats to humans. Habitat destruction and encroachment of humans into wild areas are the greatest drivers of zoonotic virus transmission from animals to humans. Some bat habitats are protected, but worldwide forests and other bat critical natural areas are dwindling. It is important to distribute accurate information about how fascinating and diverse bats are, as well as how valuable they are for for maintaining balanced ecosystems. We urgently need to conserve and provide habitats not only for bats, but also for the interconnected web of other species they interact with. Providing pesticide-free and bat-friendly gardens around our homes that

OPPOSITE TOP Land use change is the largest conservation threat to bats and most other forest-dwelling life. Scientists at STRI have expanded their research far beyond Barro Colorado Island; their focus now includes regeneration of secondary forests and forests in human-dominated landscapes. These areas are increasingly important as refugia for diverse ecological communities. OPPOSITE BOTTOM A Smithsonian scientist points out fruit bat feces full of seeds on a palm frond and discusses the critical role bats play in seed dispersal. Bats are key regenerators of the forest, especially in early successional stages.

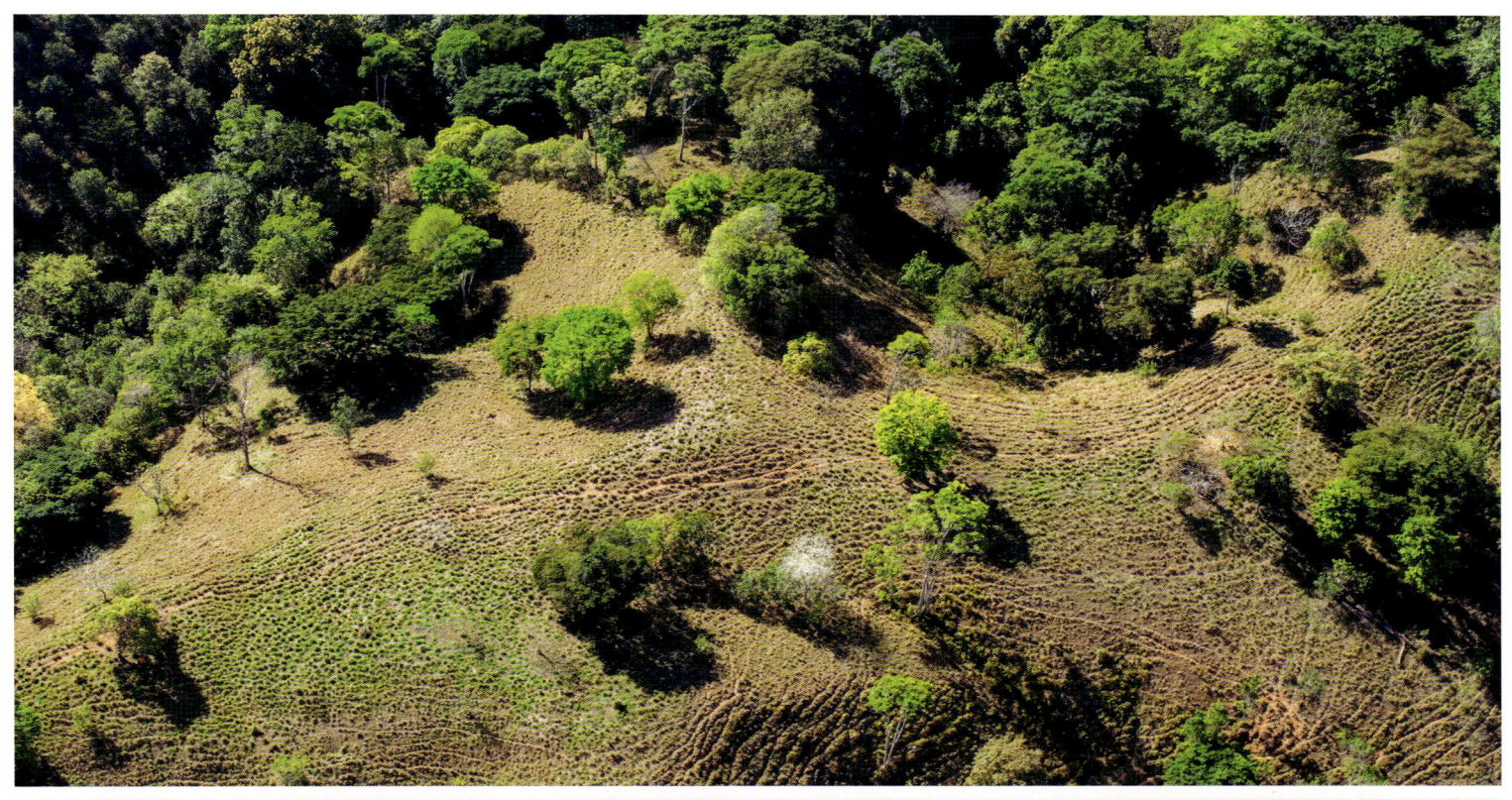

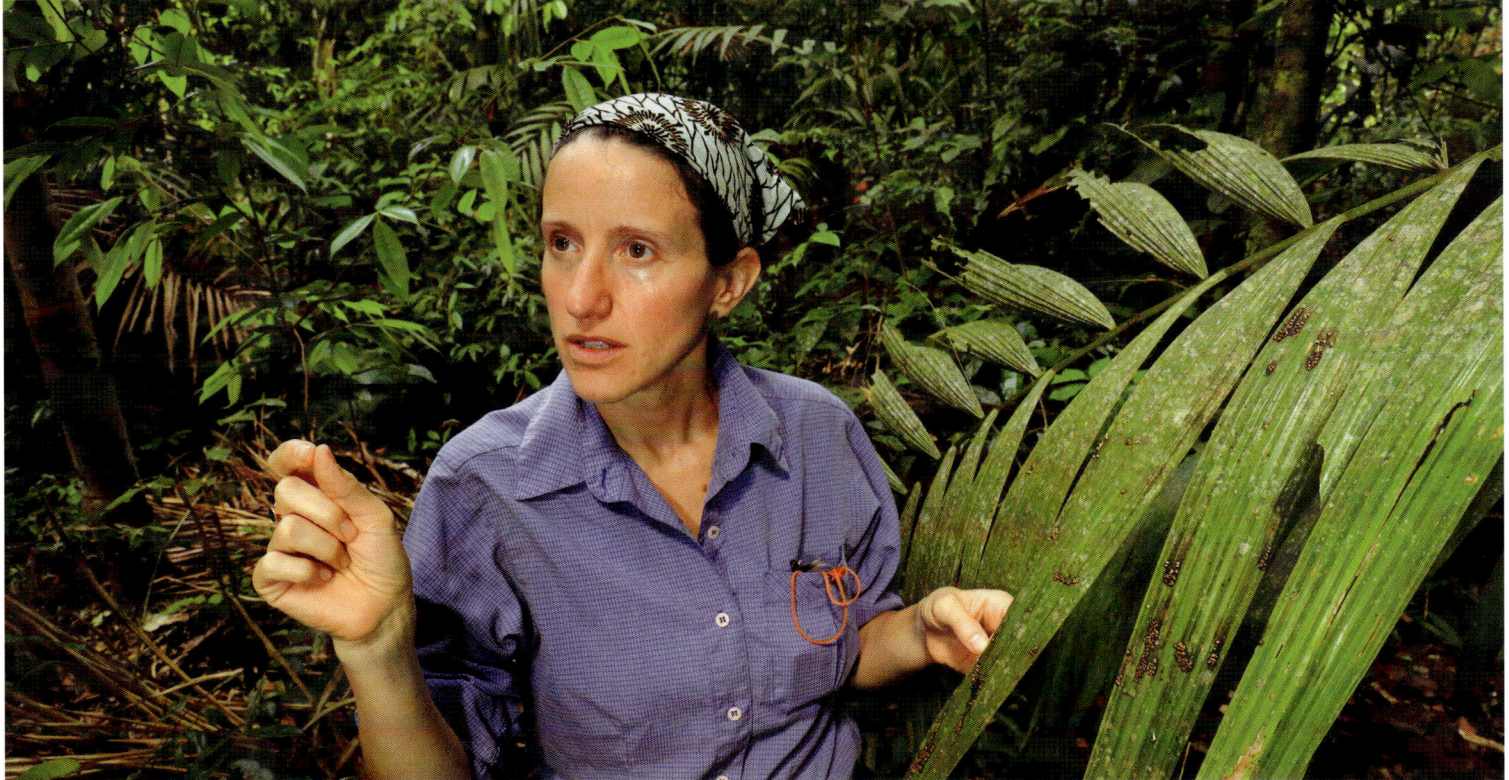

attract insect prey is one way to help bats. Known roosting sites in houses, barns, old trees, and road culverts need to be protected, and buildings can be renovated in bat-friendly ways. Cave sites need to be monitored and protected, with human extraction of guano and water harvesting limited to minimize human-bat interactions. Many NGOs are working on the complicated issues of bat conservation, and it is essential to support them so that their urgent messages are heard by lawmakers. Monetary donations help fund specific conservation actions, and local bat conservation groups are actively looking for volunteers in almost every region of the world. You can join their efforts and contribute to monitoring bat populations in buildings, mines, and forests, and ensure the safety of bat roosts. Bats need us, and we need them. The contributions you make will help protect bats and their hidden worlds for the future.

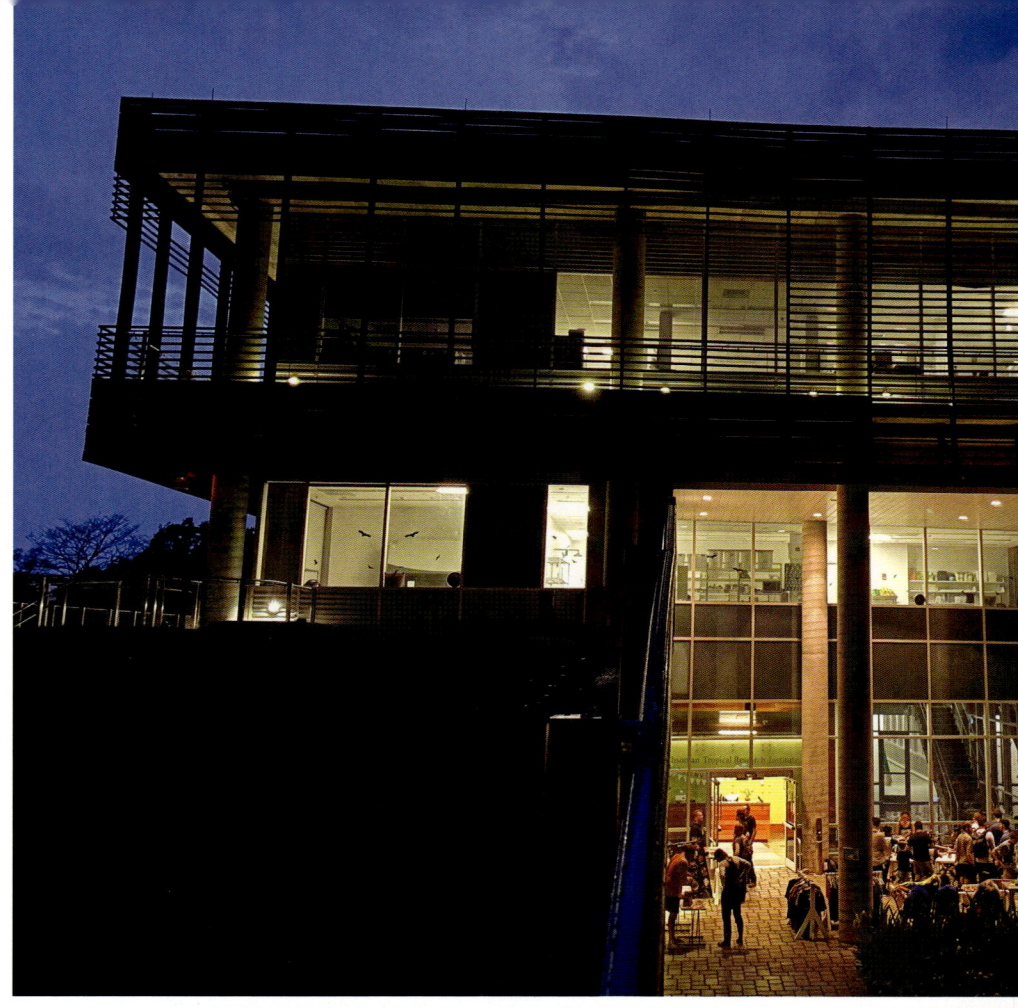

Community engagement and outreach are important ways to share the extraordinary world of bats and to inspire interest in protecting this fascinating group of mammals. The Smithsonian Tropical Research Institute holds monthly Bat Nights to offer the public a closer look at bat diversity, behavior, and ecology. Photos © Rachel A. Page (top right) and © Imran Razik (right and opposite).

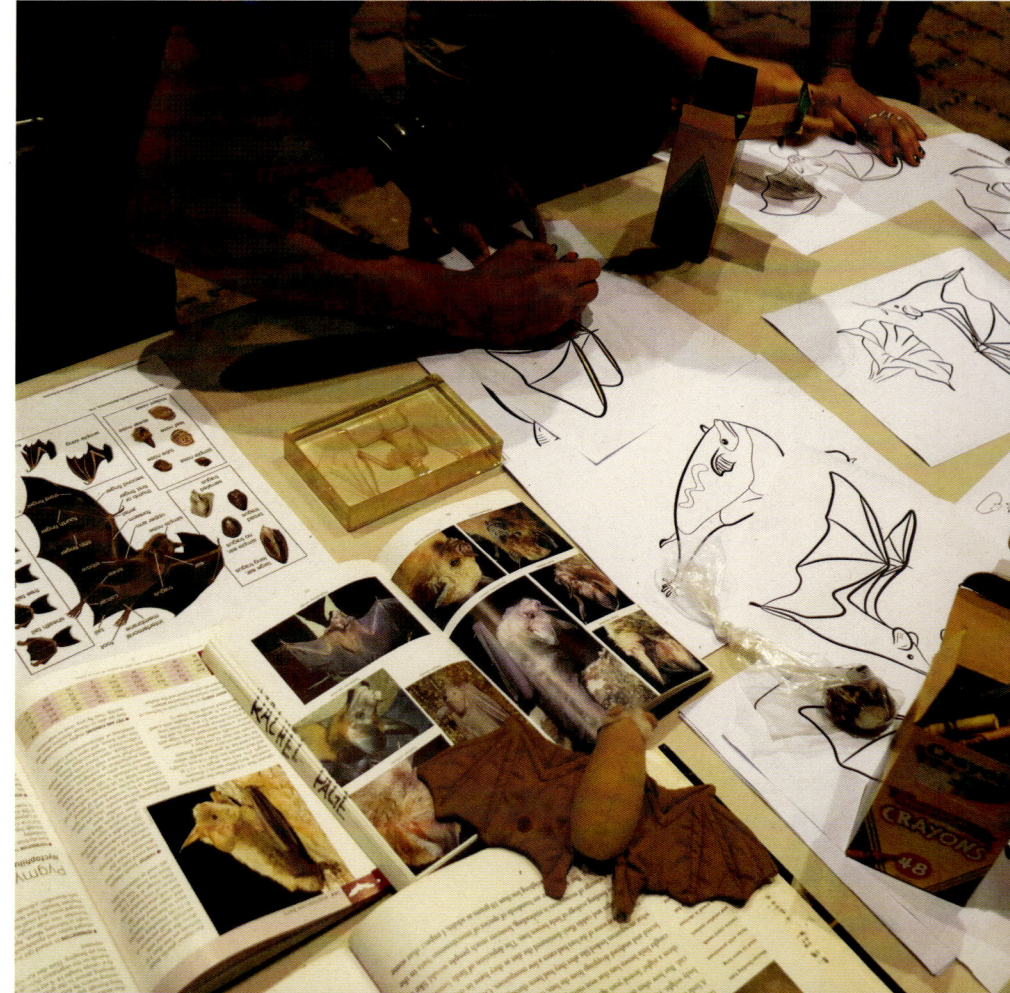

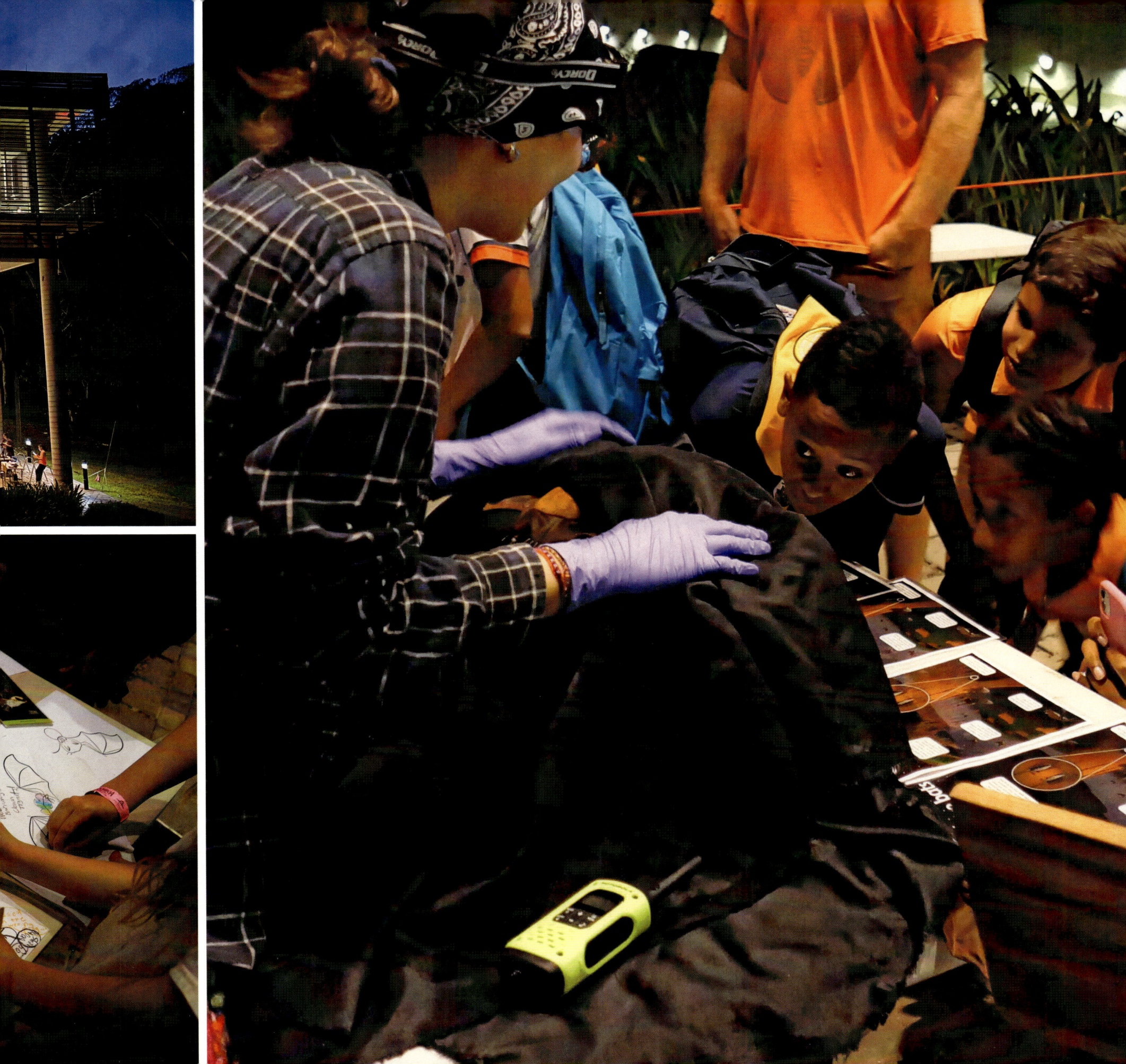

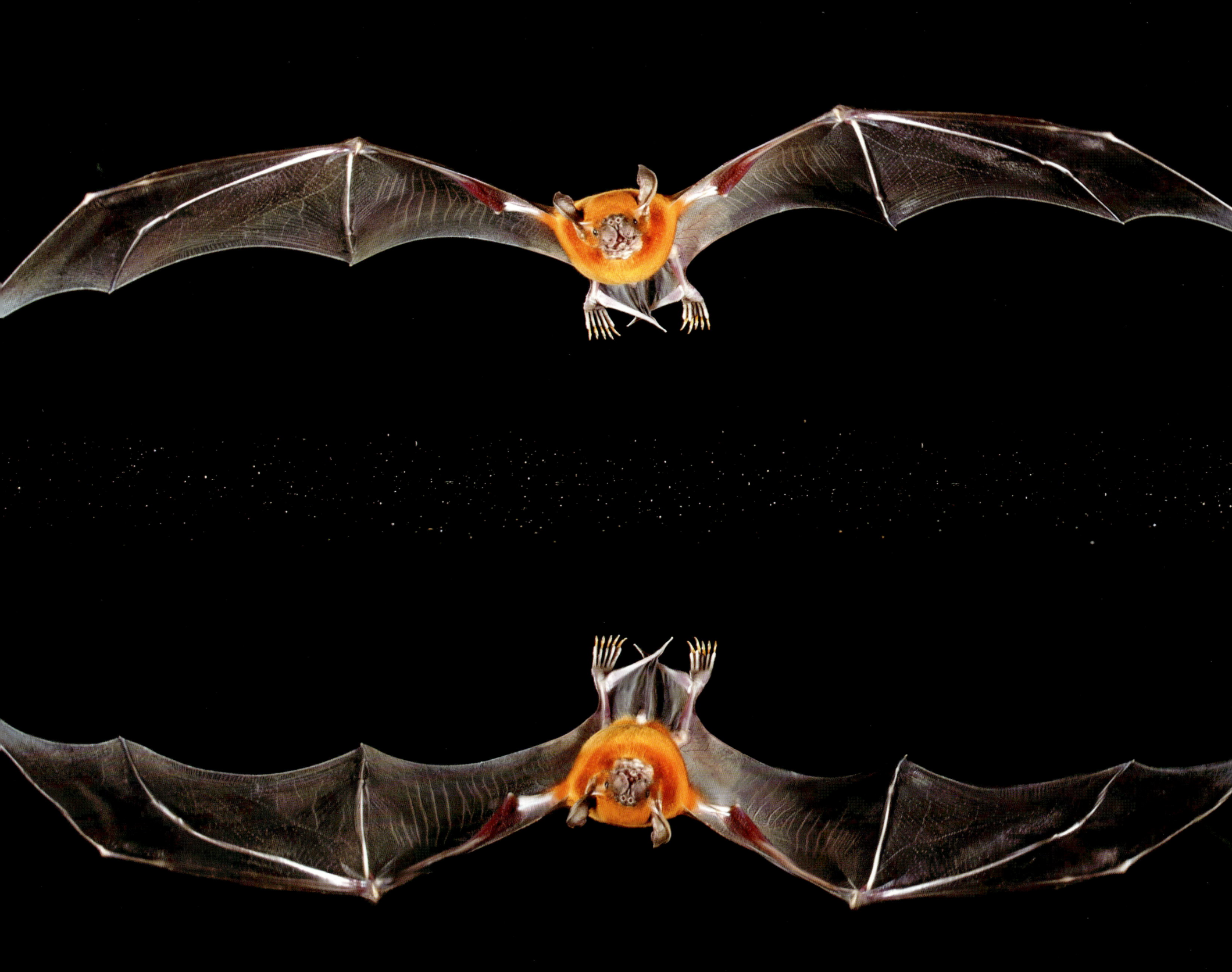

BAT CONSERVATION RESOURCES

The following is a list of some of the organizations and networks around the world dedicated to conservation and research focused on understanding and protecting the wide diversity of bats.

Australasian Bat Society (ausbats.org.au)

Bat Conservation Africa (facebook.com/groups/africasbatchampions)

Bat Conservation International (batcon.org)

Bat Conservation Trust (bats.org.uk)

BatLife Europe (batlife-europe.info)

Bats without Borders (batswithoutborders.org)

Global South Bats (globalsouthbats.org)

Global Union of Bat Diversity Networks (GBatNet) (gbatnet.org)

IUCN SSC Bat Specialist Group (iucnbsg.org)

Merlin Tuttle's Bat Conservation (merlintuttle.org)

North American Bat Conservation Alliance (NABCA) (batconservationalliance.org)

Pacific Bat Conservation Network (PacBat) (pacbat.org/)

Red Latinoamericana y del Caribe para la Conservación de los Murciélagos (RELCOM) (relcomlatinoamerica.net)

Southeast Asian Bat Conservation Research Unit (SEABCRU) (seabcru.org)

UNEP/Eurobats (eurobats.org)

Western Asia Bat Research Network (WAB-Net) (wabnet.org)

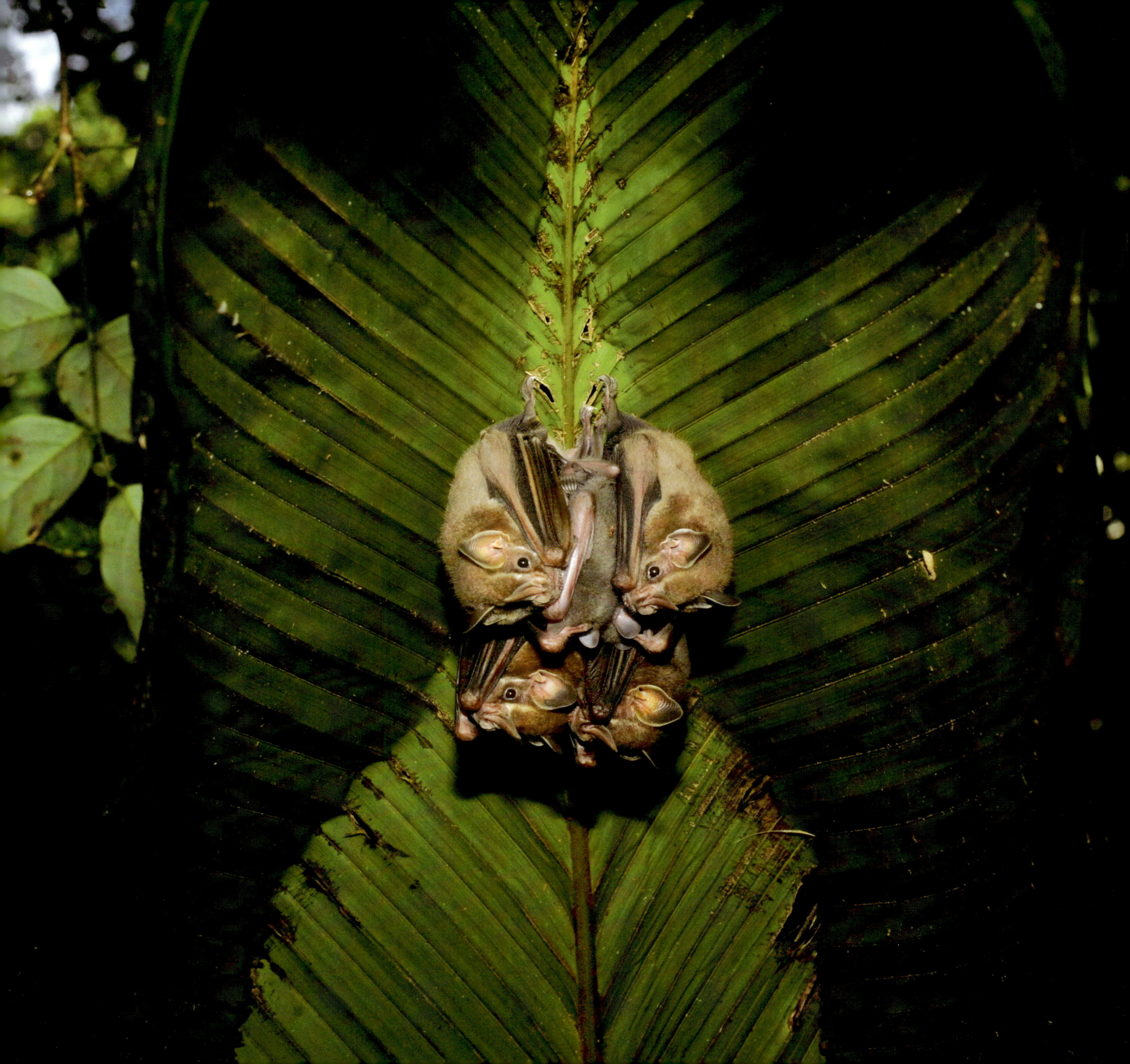

ACKNOWLEDGMENTS

We are grateful to the Smithsonian Institution, without whose support the last one hundred years of research on Barro Colorado Island would not have been possible. For valuable input on the manuscript, the authors thank numerous colleagues and friends, including Simon Bethel, Daisy Dent, May Dixon, Kirsten Jung, Mirjam Knörnschild, Sharon Martinson, Theodora Page, Laurel Symes, and Hannah ter Hofstede. We are particularly grateful to the insight provided by Mariana Muñoz-Romo.

Bats are highly mobile night creatures, notoriously challenging to capture on camera. We thank the following photographers who kindly offered their pictures to supplement the species included in this book: Gloriana Chaverri, Gregg Cohen, Inga Geipel, David Hörmann, Paul B. Jones, Karl Kugelschafter, Burton Lim, Adrià López-Baucells, José G. Martínez-Fonseca, Grant Maslowski, Manuel Sánchez Mendoza, Mariana Muñoz-Romo, Imran Razik, Clara García Sánchez, and Hubert A. Szczygieł. We are grateful to Javier Lázaro for his beautiful illustrations.

This book would not have come to fruition without the support of Linette Dutari, Jill Corcoran, Paige Towler, and the Earth Aware and Insight Editions team: Publisher Raoul Goff, Associate Publisher Roger Shaw, Publishing Director Katie Killebrew, Art Director Allister Fein, and Senior Editor John Foster.

We thank all the many BCI bat researchers who contributed to this large body of knowledge and who helped Christian Ziegler capture these mysterious and beautiful creatures with his camera lens. In particular, we are grateful to Elisabeth Kalko, whose dedicated passion for Neotropical bats has been an inspiration to us all.

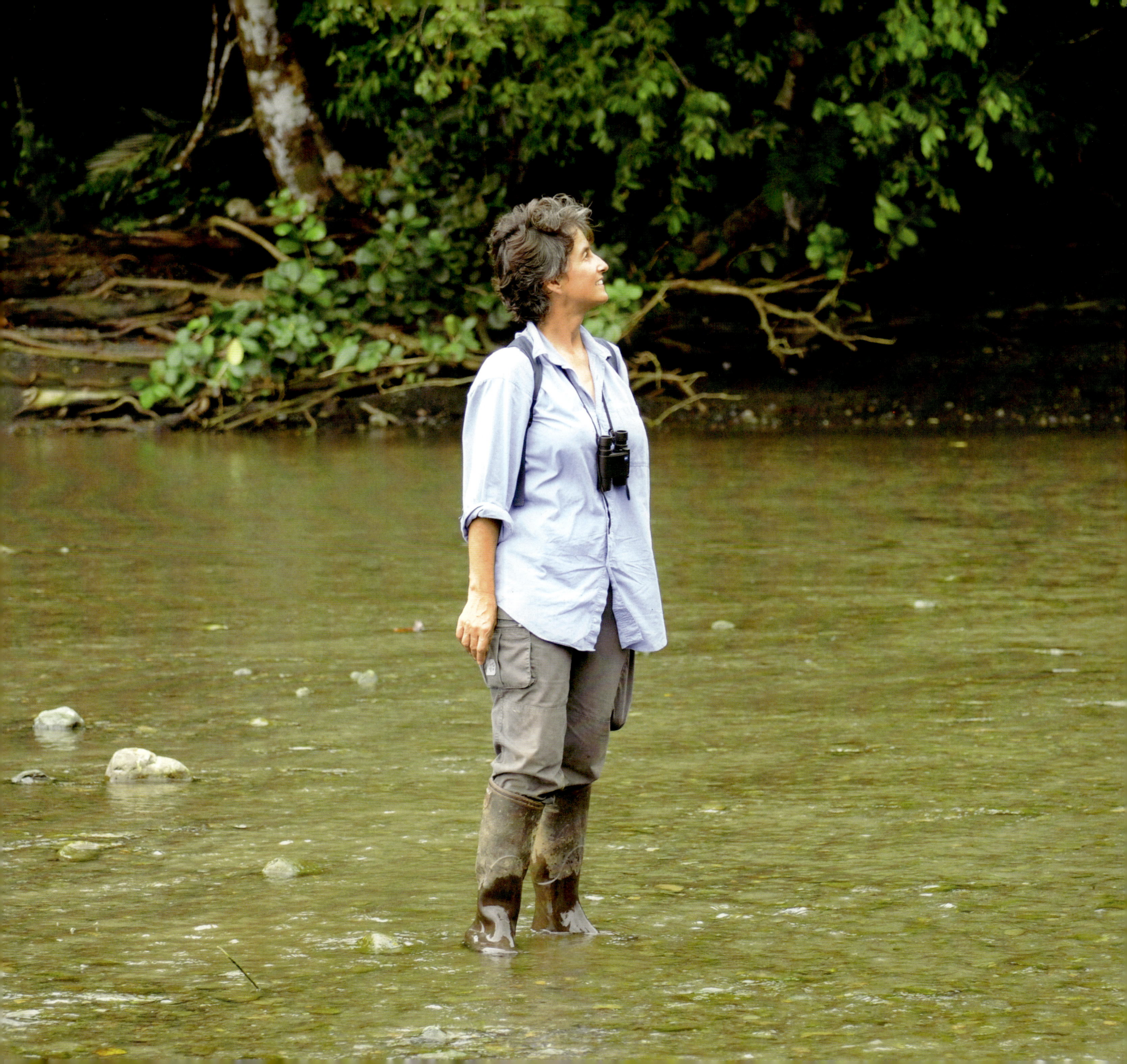

PROCEEDS FROM THE BOOK

In Elisabeth Kalko's honor, the Smithsonian Tropical Research Institute has created a memorial fellowship to support the next generation of Neotropical bat researchers. Proceeds from this book go directly to this fund. If you would like to contribute, please use the QR code below or go to stri.si.edu/donate/kalko-fund.

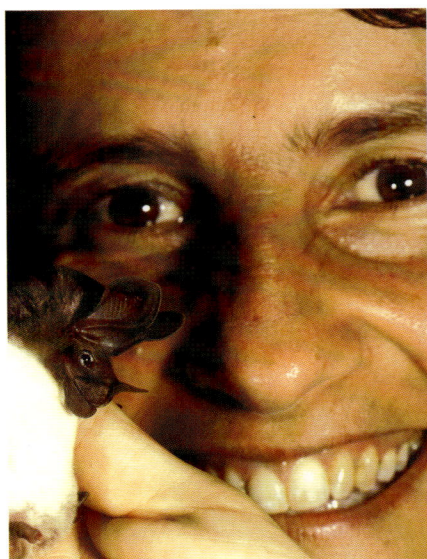
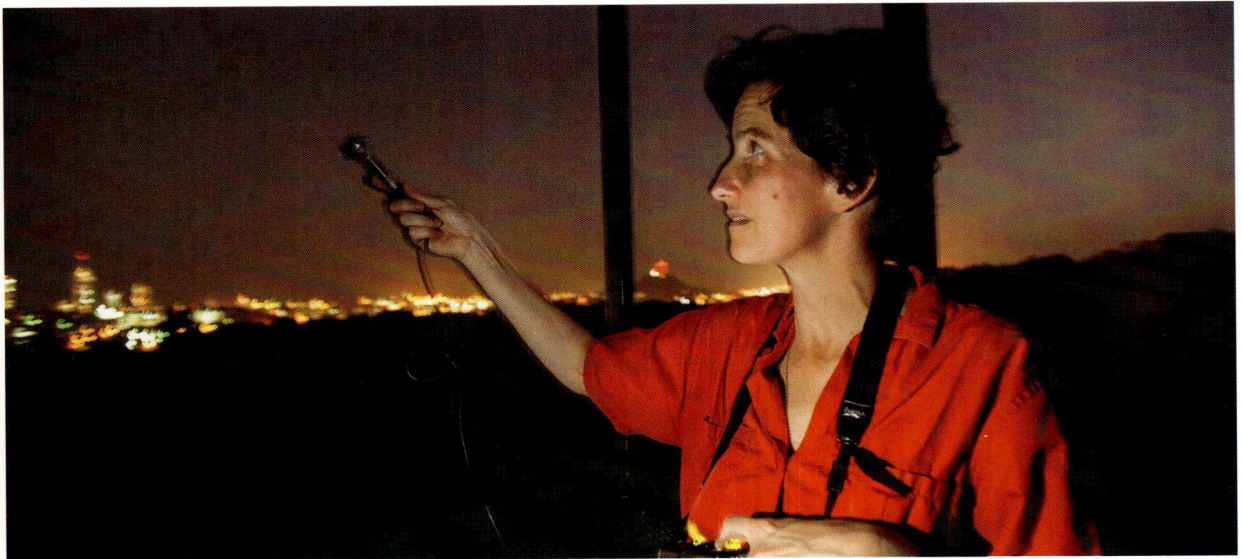

OPPOSITE AND ABOVE Elisabeth K. V. Kalko (1962–2011) in her element: wading through a tropical river, recording echolocation calls from a canopy tower, and holding Kalko's round-eared bat (*Lophostoma kalkoae*), the new species she discovered in the forests near Barro Colorado Island, Panama, which was named posthumously in her honor. The passion and dedication that Eli brought to bat research in the Neotropics continues in her students, and in their students, to this day.

INDEX

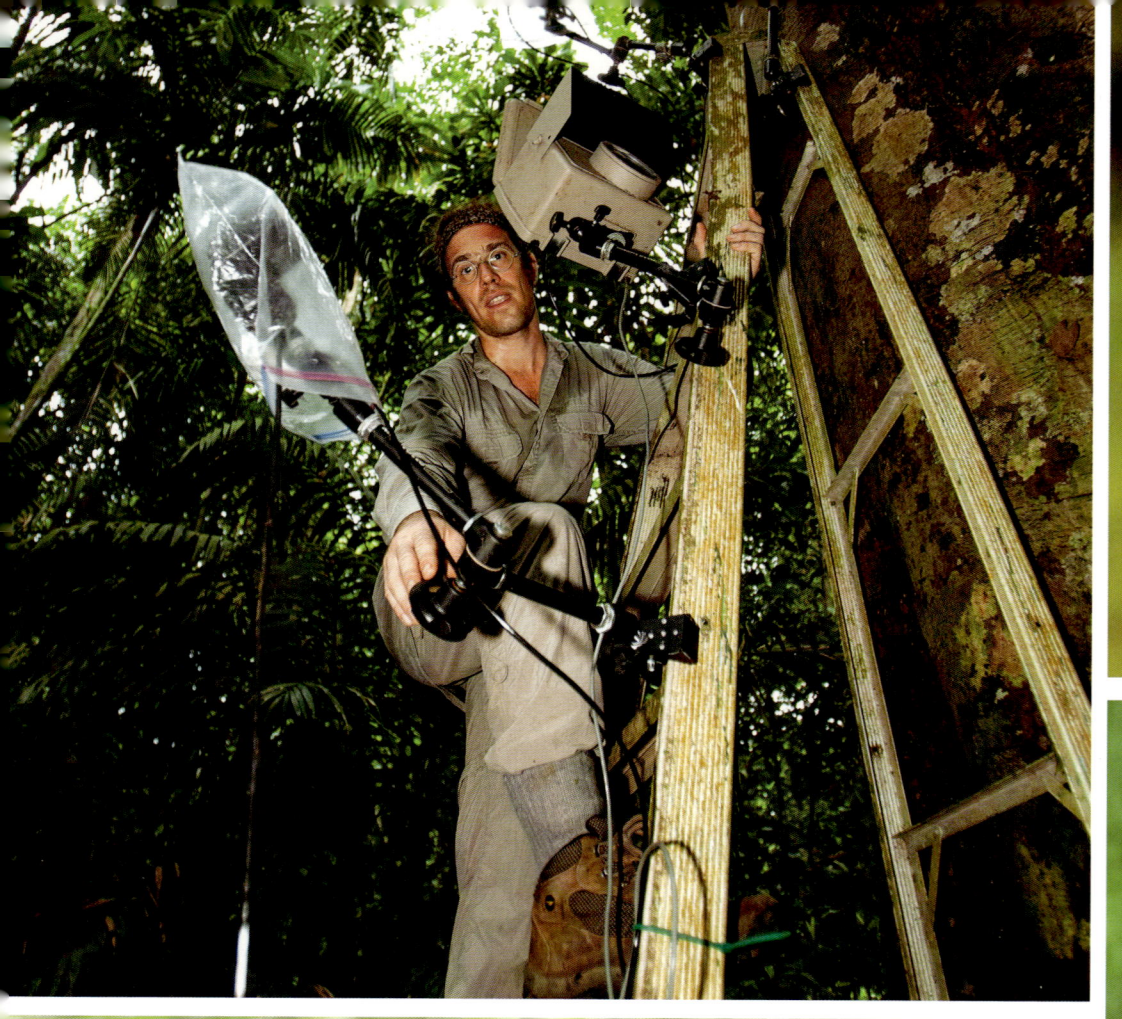
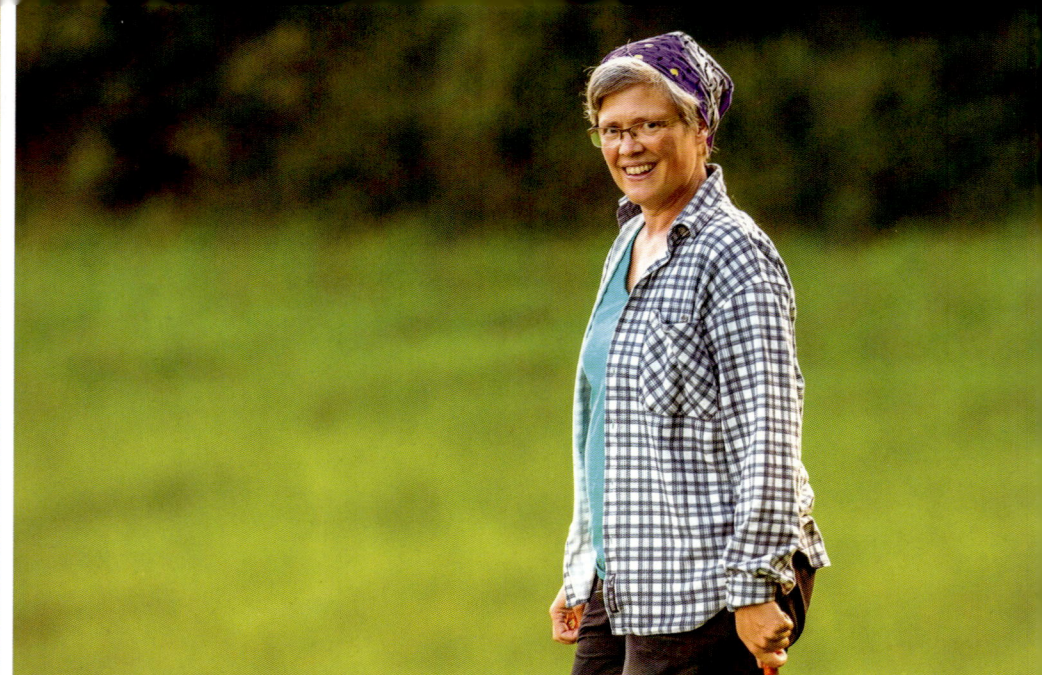
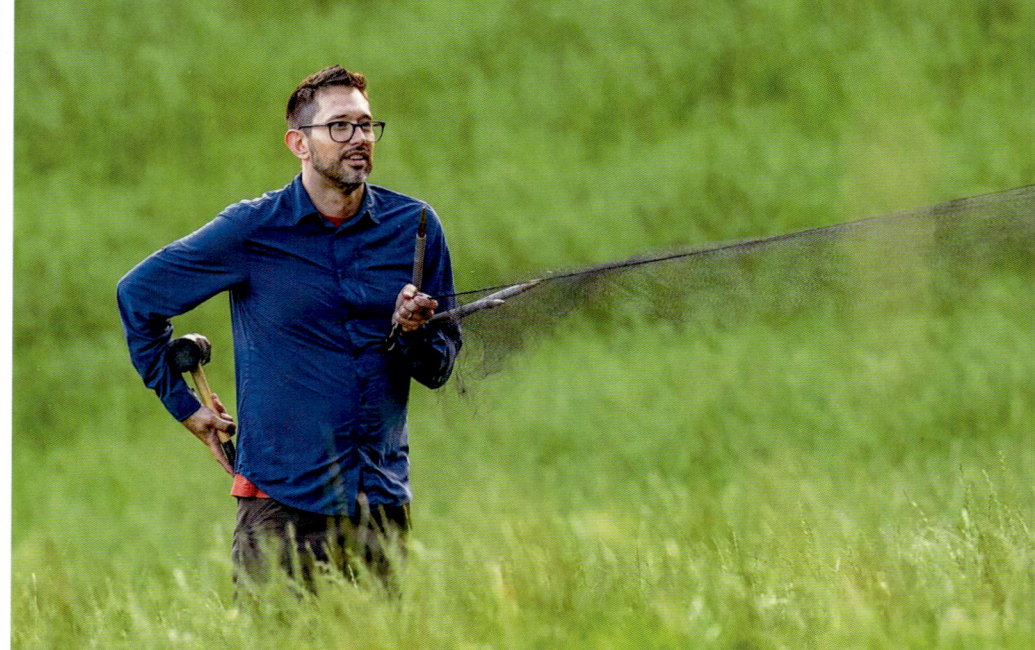
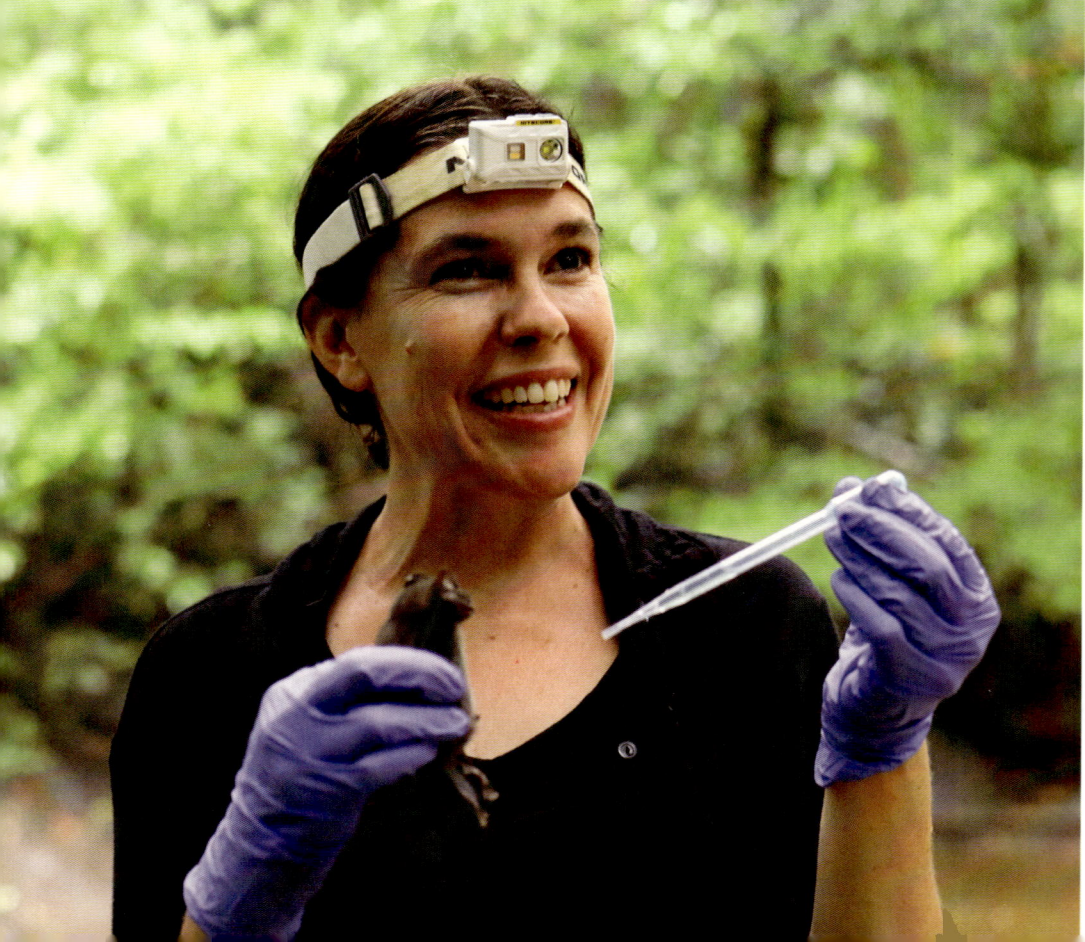
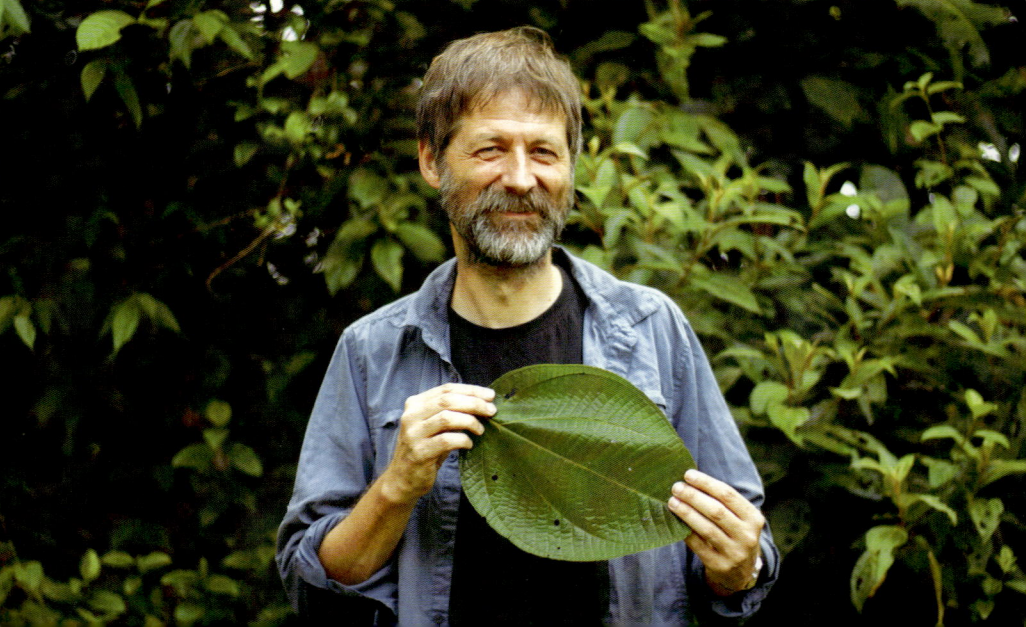

ABOUT THE PHOTOGRAPHER, AUTHORS, AND STRI

Christian Ziegler (www.christianziegler.photography) is a German photojournalist specializing in natural history and science, with a focus on tropical ecosystems. Ziegler currently works for the Max Planck Institute of Animal Behavior in Konstanz, Germany, and is a regular contributor to *National Geographic*. His work has been awarded numerous prizes in the competitions for Wildlife Photographer of the Year, European Wildlife Photographer of the Year, and the North American Nature Photography Association. A tropical ecologist by training, Ziegler has been associated with STRI for over twenty-five years since he started his own graduate research on Barro Colorado Island. He has since worked in tropical rainforests on four continents. Ziegler's aim is to highlight species and ecosystems under threat and to share their beauty and importance with the world. Ziegler splits his time between his homes on the edge of a rainforest national park in central Panama and in southern Germany, from where he starts his adventures around the world.

Dr. Rachel A. Page is a staff scientist at the Smithsonian Tropical Research Institute in Panama, where she leads the Smithsonian Bat Lab (www.noseleaf.org). She is broadly interested in animal behavior, but her focus is understanding the sensory and cognitive tools bats use to navigate their worlds and interact with each other. After completing a BA at Columbia University and a PhD at the University of Texas at Austin, Page conducted postdoctoral research as an Alexander von Humboldt fellow at the Max Planck Institute for Ornithology in Seewiesen, Germany. Page has studied bats on Barro Colorado Island and the surrounding areas for over two decades. She has a passion for understanding rich, tropical ecosystems and the myriad species interactions they encompass. In addition to conducting her own research, Page mentors a large group of students. Page lives at the edge of the rainforest in Gamboa, Panama.

Dr. Dina K. N. Dechmann is a group leader at the Max Planck Institute of Animal Behavior (www.ab.mpg.de/dechmann) and a research associate at the Smithsonian Tropical Research Institute. An evolutionary ecologist by training, her main research interest is how animals adapt to fluctuations in the resources upon which they depend. She is fascinated by how tiny mammals with fast metabolisms, such as bats and shrews, adapt their morphology, physiology, and behavior to deal with the bottlenecks created by changes in the food landscape. She works in ecosystems across the world, but since her first visit to BCI in 2000, she has been struck by the diversity of the tropical bat community, which remains a cornerstone of her research. She received her master's degree at the Swiss Federal Institute of Technology (ETH) in Zurich followed by a PhD at the University of Zurich, and was a postdoctoral fellow at the Leibniz Institute for Zoo and Wildlife Research in Berlin. Since she was hired by Max Planck in 2009, she has had the good fortune to supervise a group of brilliant young minds, several of whom now run their own projects in Panama. Dechmann lives in the medieval town of Stein am Rhein in Switzerland.

Dr. M. Teague O'Mara is the Director of Conservation Evidence at Bat Conservation International, where he works on data-driven strategies for the conservation of global bat populations. O'Mara has studied animal behavior, movement, and physiology across the globe, with an emphasis on bats in Panama. He is a research associate at the Smithsonian Tropical Research Institute and the Max Planck Institute of Animal Behavior, and an adjunct professor at Southeastern Louisiana University. He received his PhD from Arizona State University studying lemur development and social behavior, and then switched to research with bats during postdoctoral work at the Smithsonian Tropical Research Institute, the University of Konstanz, and the Max Planck Institute of Animal Behavior in Germany. He lives in Baton Rouge, Louisiana.

Dr. Marco Tschapka is a professor at the University of Ulm in Germany and a research associate at the Smithsonian Tropical Research Institute. His research addresses bat-plant interactions, with a focus on the ecology of Neotropical nectar-feeding bats, their adaptations to nectarivory, and the plants they visit and pollinate. Tschapka received his master's degree and PhD with Dr. Otto von Helversen at the University of Erlangen, Germany, on Neotropical flower-visiting bats. Together with his collaborators and students, he has worked on bat projects across Latin America, including in Mexico, Costa Rica, Ecuador, Peru, Brazil, and Panama. He lives in Ulm, Germany.

ABOUT THE SMITHSONIAN TROPICAL RESEARCH INSTITUTE

The **Smithsonian Tropical Research Institute**, headquartered in Panama City, Panama, is a unit of the Smithsonian Institution in Washington, D.C., USA. The institute furthers the understanding of tropical biodiversity and its importance to human welfare, trains students to conduct research in the tropics, and promotes conservation by increasing public awareness of the beauty and importance of tropical ecosystems and their cultures.

CLOCKWISE FROM TOP LEFT Christian Ziegler; Dr. Dina K.N. Dechmann; Dr. M. Teague O'Mara (photos © Christian Ziegler); Dr. Marco Tschapka (photo © Clara García Sánchez); and Dr. Rachel A. Page (photo © Imran Razik).

THE BATS OF
BARRO COLORADO ISLAND

The bats of Barro Colorado Island are highly diverse and make their livings in widely different ways. Some predators dart swiftly through the air in pursuit of flying insects, while others fly slowly through the understory and scan leaf surfaces for sleeping insects or tiny lizards. Some species listen with giant ears to the sounds of the forest, locating singing katydids or even male frogs serenading mates from tiny puddles, only to discover too late that their love song was also heard by a giant eavesdropping predator. The slightest rustling sound can betray a mouse running over dry leaves on the forest floor to the larger of these predatory species. Not even fish in the water are safe from nocturnal hunters that swoop down and grasp their slippery prey with sharp claws. More peaceful are the vegetarians; bats attracted to a multitude of fruit trees that rely on these flying frugivores to distribute their seeds. Fruit-eating bats come in all sizes and feed on matching fruits that range from the size of a blueberry up to that of a large plum. Nocturnal flowers depend on the pollination services of bats, which become covered in pollen while licking nectar deep within flowers, with long tongues that can extend to nearly the length of their bodies.

The members of the island's night shift come from a variety of taxonomic origins, with evolution shaping each species to be well adapted to its specific lifestyle. Of the eight families of bats present on BCI, seven are exclusively predators, while the last and largest family, the Neotropical leaf-nosed bats (Phyllostomidae), exhibits one of the most amazing adaptive radiations of all mammals and contains not only predators, but also frugivores, nectarivores, and even the infamous vampire bats that feed only on the blood of vertebrates.

The following pages showcase the amazing morphological diversity of this nocturnal cast of characters on Barro Colorado Island. Portraits of BCI's seventy-six bat species are color coded to indicate their taxonomic group, while their preferred and primary diet is indicated by diet icons. Photo credits are indicated by the following symbols: ⋏ Gloriana Chaverri • David Hörmann ✪ Burton Lim, ◆ Adrià López-Baucells ◉ José G. Martínez-Fonseca ❖ Mariana Muñoz-Romo ❇ Marco Tschapka

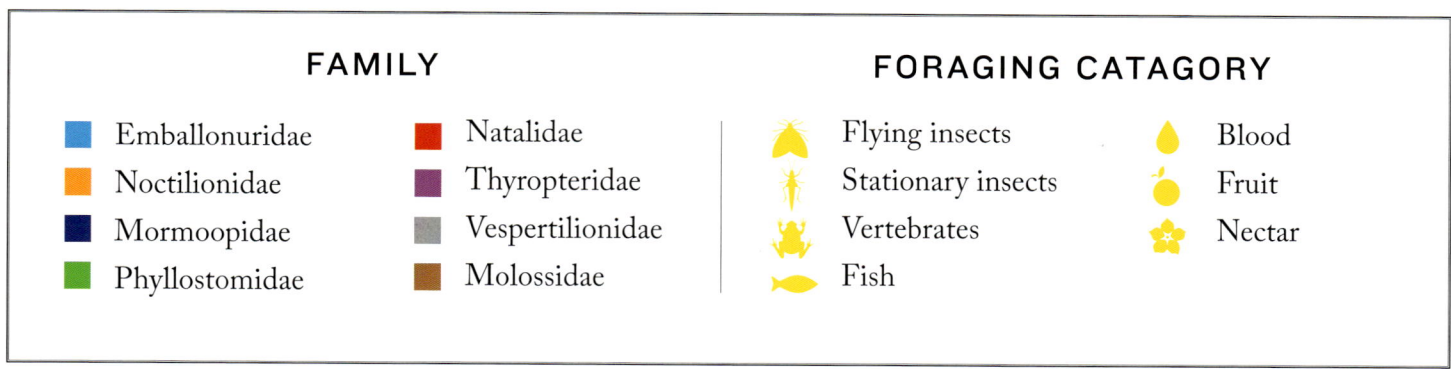

FAMILY		FORAGING CATAGORY	
■ Emballonuridae	■ Natalidae	⋏ Flying insects	● Blood
■ Noctilionidae	■ Thyropteridae	Stationary insects	● Fruit
■ Mormoopidae	■ Vespertilionidae	Vertebrates	❋ Nectar
■ Phyllostomidae	■ Molossidae	Fish	

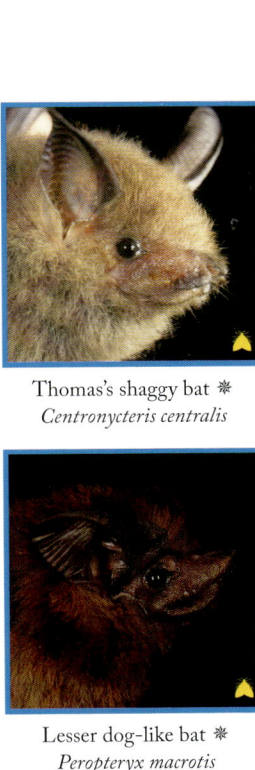

Thomas's shaggy bat ✳
Centronycteris centralis

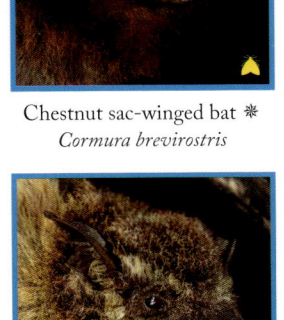

Chestnut sac-winged bat ✳
Cormura brevirostris

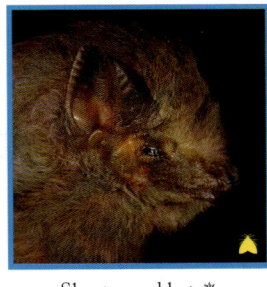

Short-eared bat ✳
Cyttarops alecto

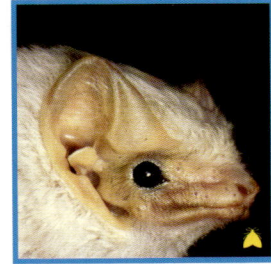

Northern ghost bat ❖
Diclidurus albus

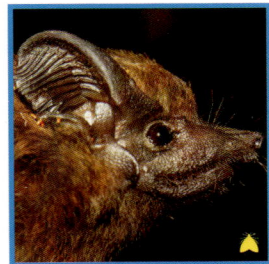

Greater dog-like bat ◆
Peropteryx kappleri

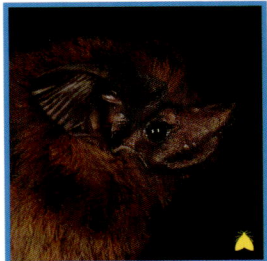

Lesser dog-like bat ✳
Peropteryx macrotis

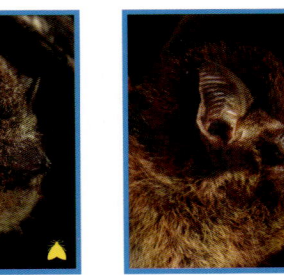

Proboscis bat ◆
Rhynchonycteris naso

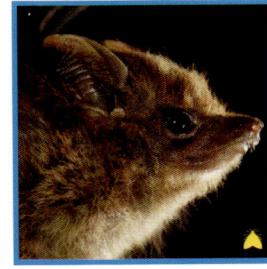

Greater sac-winged bat ✳
Saccopteryx bilineata

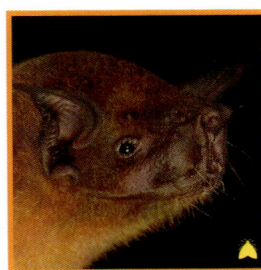

Lesser sac-winged bat ◆
Saccopteryx leptura

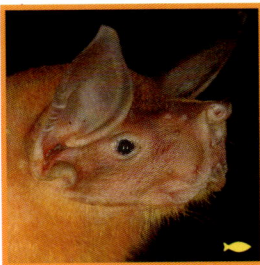

Lesser bulldog bat ✳
Noctilio albiventris

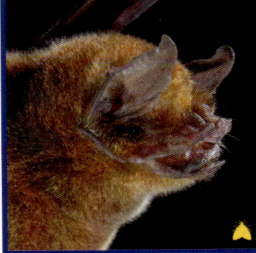

Greater bulldog or fishing bat ✳
Noctilio leporinus

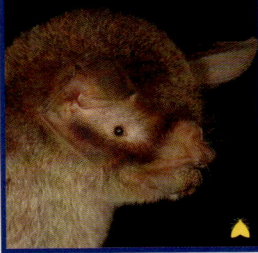

Big naked-backed bat ◆
Pteronotus gymnonotus

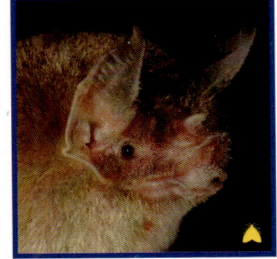

Parnell's mustached bat ✳
Pteronotus parnellii

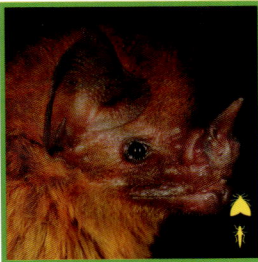

Wagner's mustached bat ✳
Pteronotus personatus

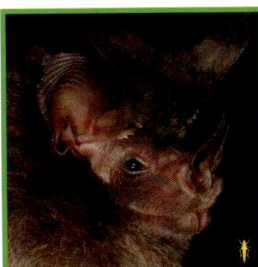

Orange-throated bat ✳
Lampronycteris brachyotis

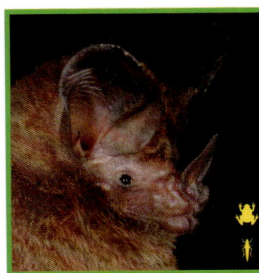

Hairy-big-eared bat ✳
Micronycteris hirsuta

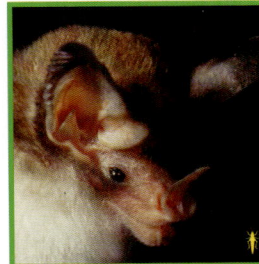

Common big-eared bat ✳
Micronycteris microtis

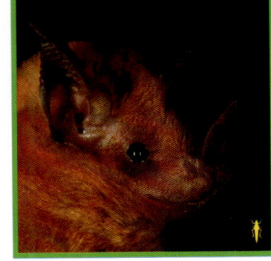

Schmidt's big-eared bat ✳
Micronycteris schmidtorum

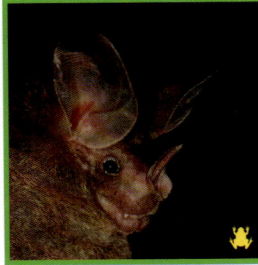

Niceforo's big-eared bat ✳
Trinycteris nicefori

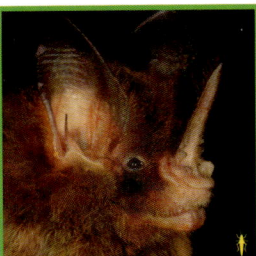

Big-eared woolly bat ✳
Chrotopterus auritus

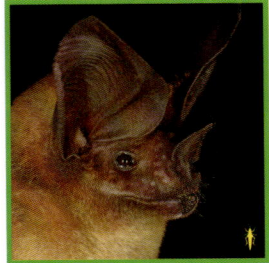

Striped hairy-nosed bat ✳
Gardnerycteris crenulata

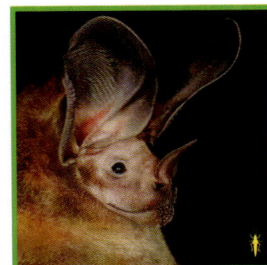

Pygmy round-eared bat ✳
Lophostoma brasiliense

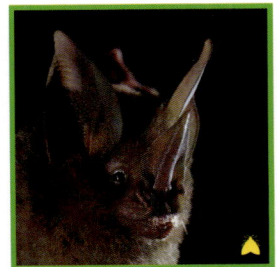

White-throated round-eared bat ✳
Lophostoma silvicolum

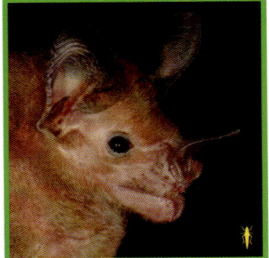

Long-legged bat ◆
Macrophyllum macrophyllum

Pale-faced bat ✳
Phylloderma stenops

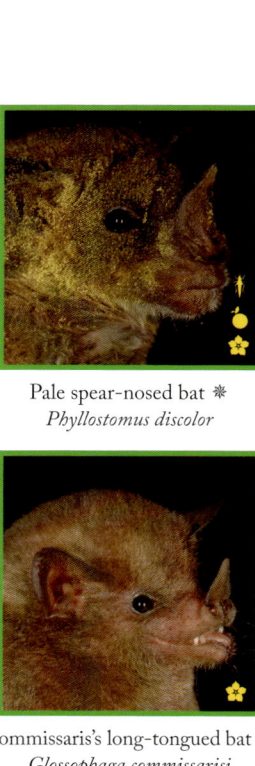

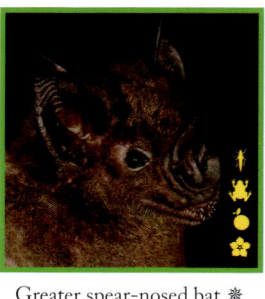

Pale spear-nosed bat ✳
Phyllostomus discolor

Greater spear-nosed bat ✳
Phyllostomus hastatus

Stripe-headed round-eared bat ✳
Tonatia saurophila

Fringe-lipped bat ✳
Trachops cirrhosus

Spectral bat ✳
Vampyrum spectrum

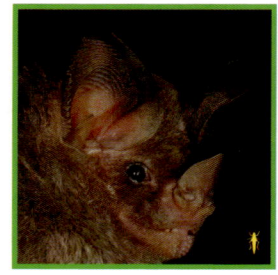
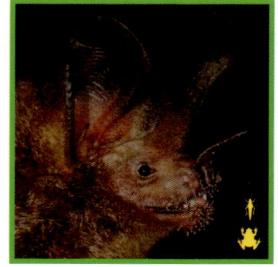
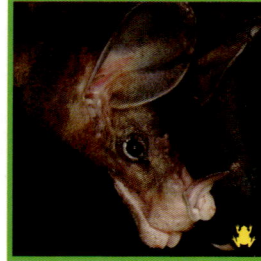

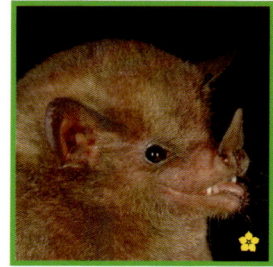
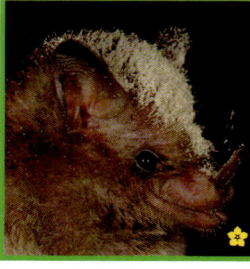
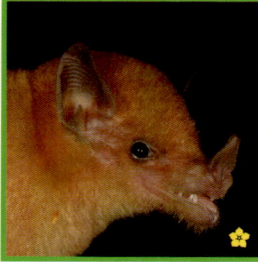
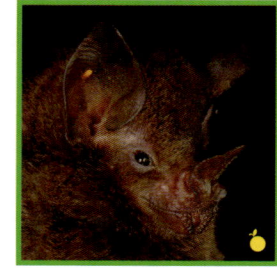
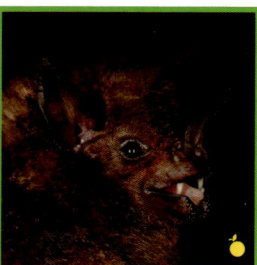

Commissaris's long-tongued bat ✳
Glossophaga commissarisi

Pallas's long-tongued bat ✳
Glossophaga soricina

Orange nectar bat ✳
Lonchophylla robusta

Silky short-tailed bat ✳
Carollia brevicauda

Chestnut short-tailed bat ✳
Carollia castanea

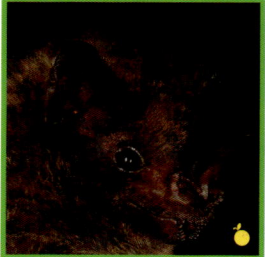
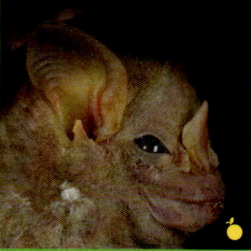
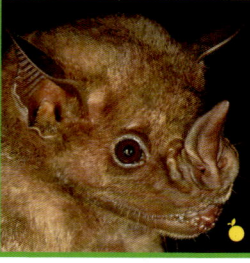
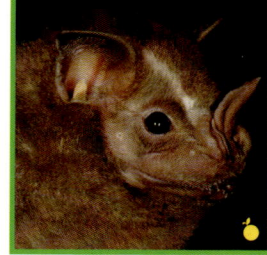
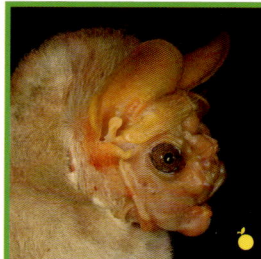

Seba's short-tailed bat ✳
Carollia perspicillata

Little white-shouldered bat ✿
Ametrida centurio

Jamaican fruit-eating bat ✳
Artibeus jamaicensis

Great fruit-eating bat ✳
Artibeus lituratus

Wrinkle-faced bat ✳
Centurio senex

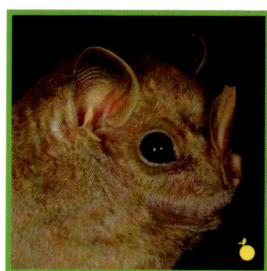
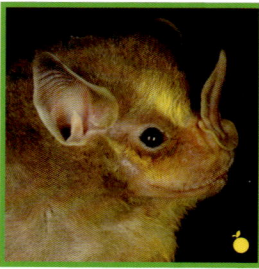
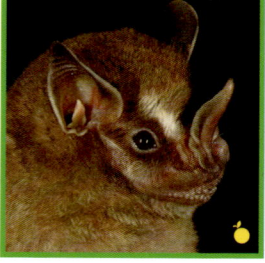
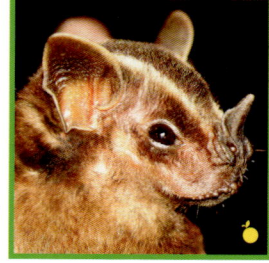
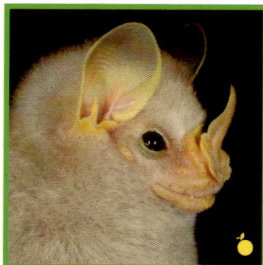

Hairy big-eyed bat ✳
Chiroderma villosum

Pygmy fruit-eating bat ✳
Dermanura phaeotis

Thomas' fruit-eating bat ✳
Dermanura watsoni

Velvety fruit-eating bat •
Enchisthenes hartii

MacConnell's bat ✳
Mesophylla macconnelli

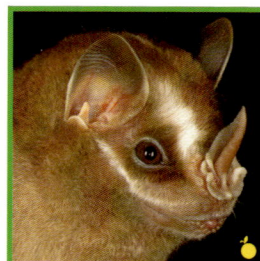
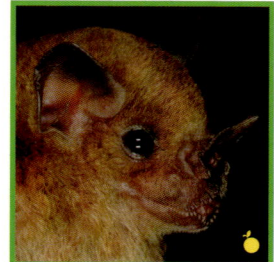
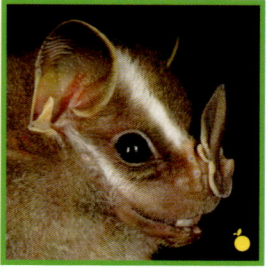
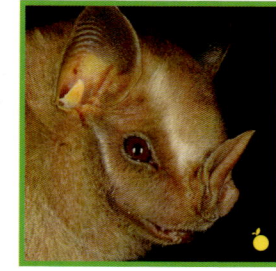
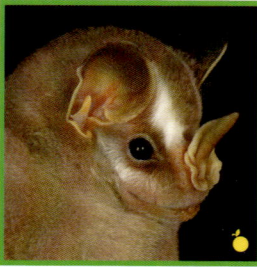

Heller's broad-nosed bat ✳
Platyrrhinus helleri

Luis' yellow-shouldered bat ✳
Sturnira luisi

Peters' tent-making bat ✳
Uroderma bilobatum

Brown tent-making bat ✳
Uroderma magnirostrum

Striped yellow-eared bat ✳
Vampyriscus nymphaea

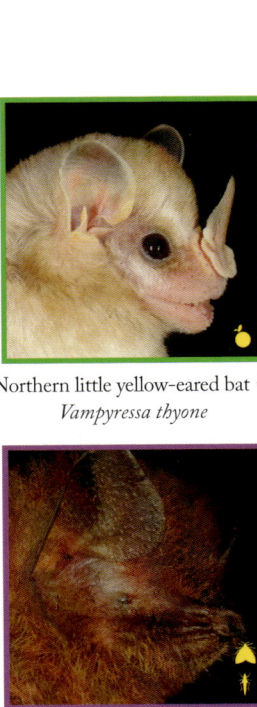

Northern little yellow-eared bat ✳
Vampyressa thyone

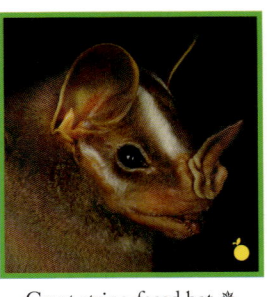

Great stripe-faced bat ✳
Vampyrodes caraccioli

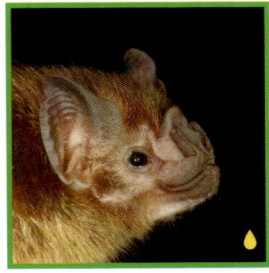

Common vampire bat ✳
Desmodus rotundus

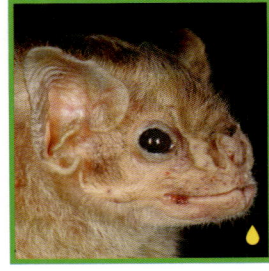

White-winged vampire bat ✳
Diaemus youngii

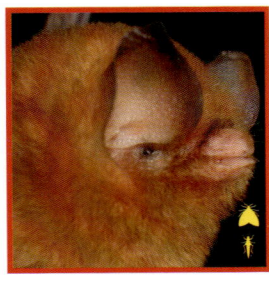

Mexican funnel-eared bat ✳
Natalus stramineus

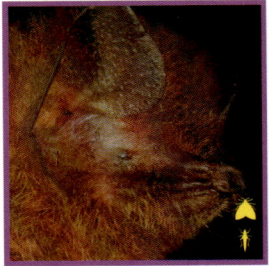

Peters's disk-winged bat ✳
Thyroptera discifera

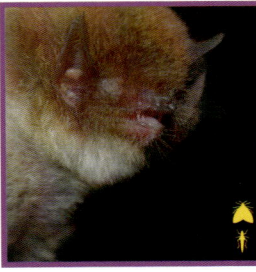

Spix's disk-winged bat ⏶
Thyroptera tricolor

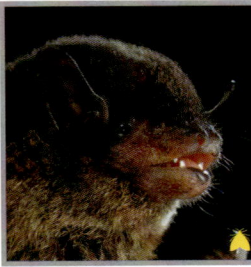

Brazilian brown bat ◆
Eptesicus brasiliensis

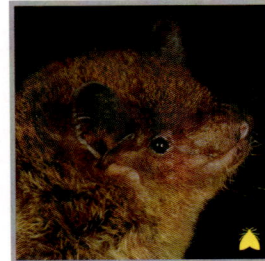

Argentine brown bat ✳
Eptesicus furinalis

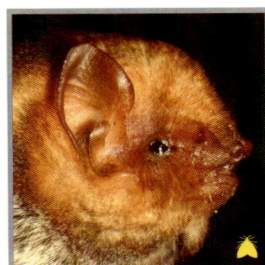

Western red bat ❖
Lasiurus blossevillii

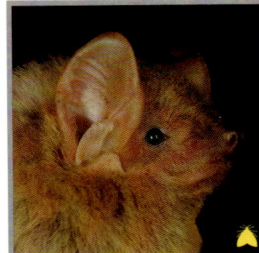

Southern yellow bat ✳
Lasiurus ega

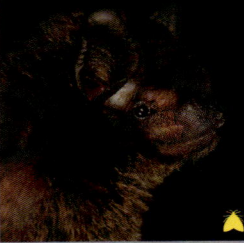

Silver-tipped myotis ✳
Myotis albescens

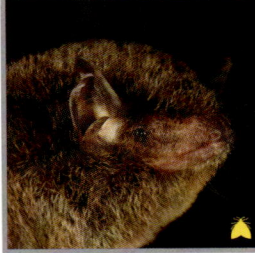

Black myotis ✳
Myotis nigricans

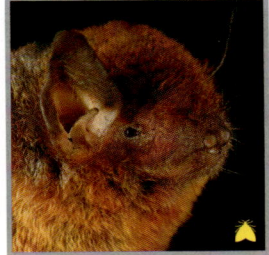

Riparian myotis ✳
Myotis riparius

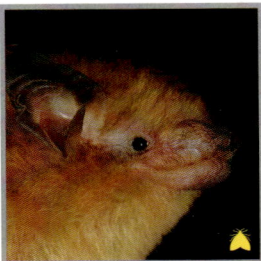

Black-winged little yellow bat ✳
Rhogeessa tumida

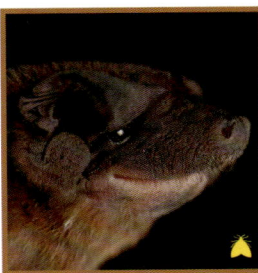

Greenhall's dog-faced bat ◎
Cynomops greenhalli

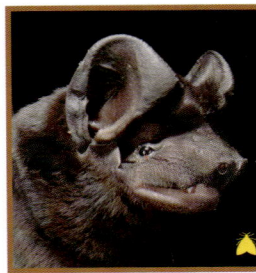

Black bonneted bat ✿
Eumops auripendulus

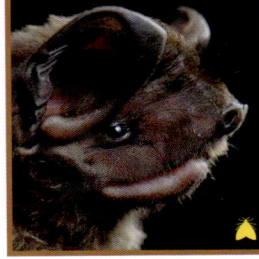

Wagner's mastiff bat ◎
Eumops glaucinus

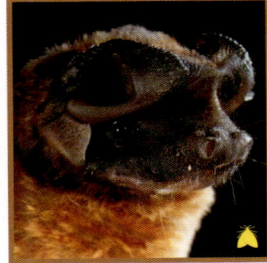

Sanborn's bonneted bat ◆
Eumops hansae

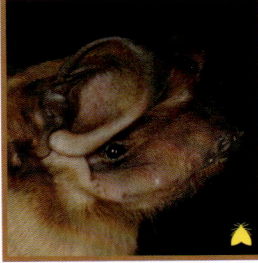

Underwood's bonneted bat ✳
Eumops underwoodi

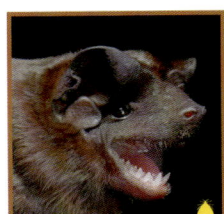

Coiban mastiff bat ✿
Molossus coibensis

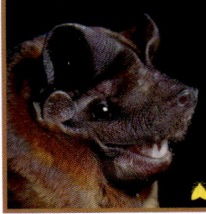

Bonda mastiff bat ◎
Molossus currentium

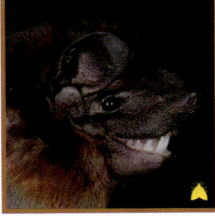

Pallas's mastiff bat ✳
Molossus molossus

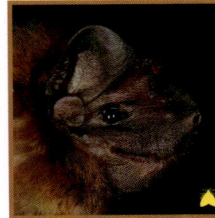

Sinaloan mastiff bat ✳
Molossus sinaloae

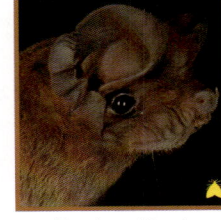

Broad-eared bat ✳
Nyctinomops laticaudatus

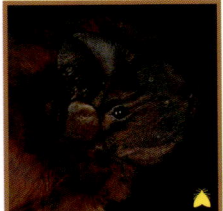

Big-crested mastiff bat ✳
Promops centralis

EARTH AWARE

An Imprint of MandalaEarth
PO Box 3088
San Rafael, CA 94912
www.MandalaEarth.com

Find us on Facebook: www.facebook.com/MandalaEarth
Follow us on Twitter: @MandalaEarth

Publisher Raoul Goff
Associate Publisher Roger Shaw
Publishing Director Katie Killebrew
Senior Editor John Foster
Editorial Assistant Amanda Nelson
VP, Creative Director Chrissy Kwasnik
Art Director Allister Fein
VP Manufacturing Alix Nicholaeff
Sr Production Manager Joshua Smith
Sr Production Manager, Subsidiary Rights Lina s Palma-Temena

MandalaEarth would also like to thank Bob Cooper for copyediting, Margaret Parrish for proofreading , and Kevin Broccoli for indexing.

For Smithsonian Enterprises:
Licensing Coordinator Avery Naughton,
Editorial Lead Paige Towler,
Senior Director, Licensed Publishing Jill Corcoran
Vice President of New Business and Licensing Brigid Corcoran
President Carol LeBlanc
Office of Communications Smithsonian Tropical Research Institute

ISBN: 979-8-88762-039-8

Manufactured in China by Insight Editions
10 9 8 7 6 5 4 3 2 1

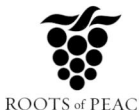

ROOTS of PEACE   🌲 REPLANTED PAPER

Insight Editions, in association with Roots of Peace, will plant two trees for each tree used in the manufacturing of this book. Roots of Peace is an internationally renowned humanitarian organization dedicated to eradicating land mines worldwide and converting war-torn lands into productive farms and wildlife habitats. Roots of Peace will plant two million fruit and nut trees in Afghanistan and provide farmers there with the skills and support necessary for sustainable land use.

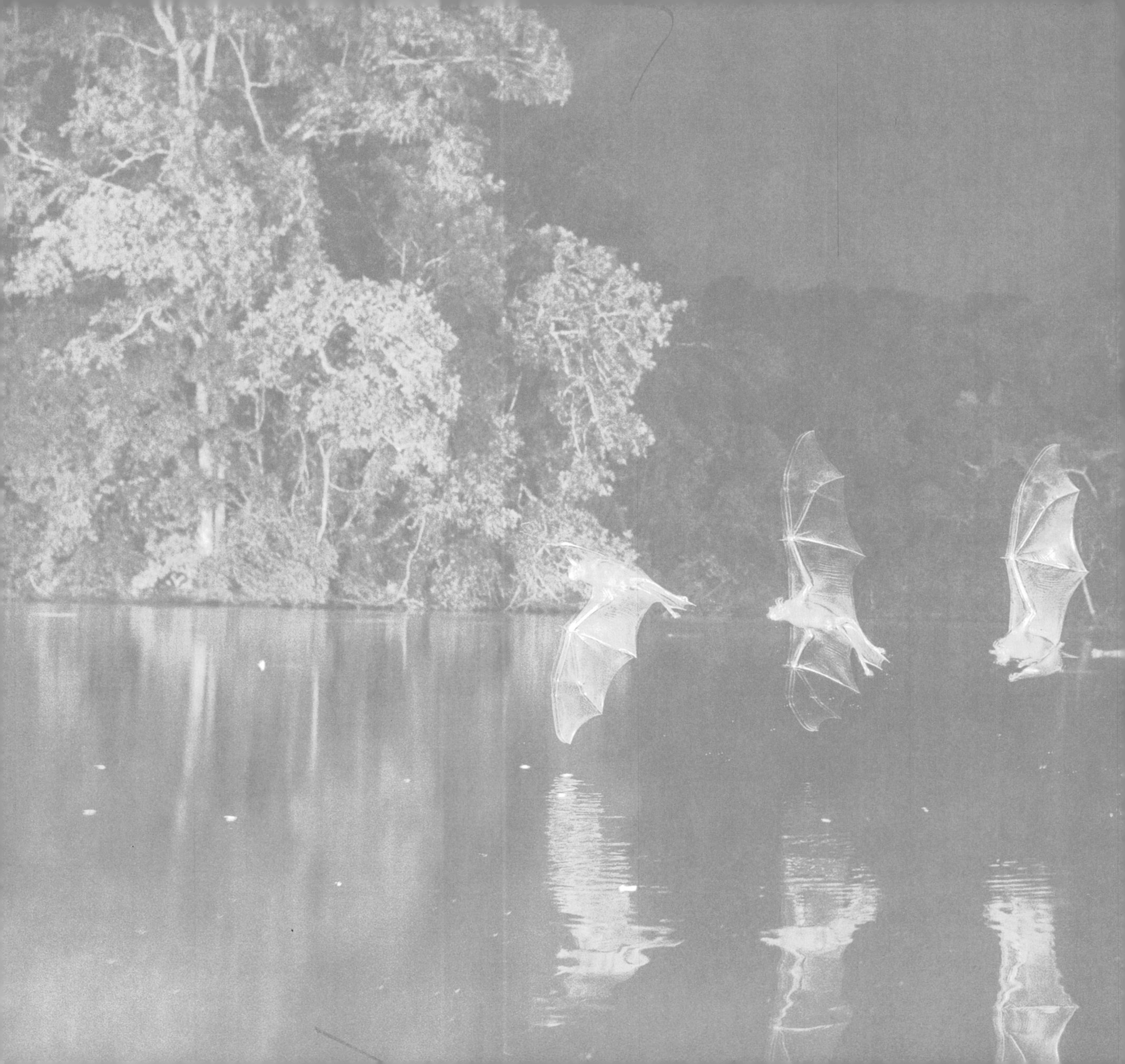